The Fundamentals of
Motor Vehicle Electrical Systems

The Fundamentals of Motor Vehicle Electrical Systems

JOHN HARTLEY
IEng MIEIE

Longman Scientific & Technical

Longman Scientific & Technical,
Longman Group UK Limited,
Longman House, Burnt Mill, Harlow,
Essex CM20 2JE, England
and Associated Companies throughout the world.

First published 1992

ISBN 0 582 08699 X

British Library Cataloguing in Publication Data
A catalogue record for this book is available from the British Library

Disc conversion in 10/12 point Times by 8

Printed in Hong Kong
WC/01

Contents

Preface

This book provides a comprehensive coverage of the basic electrical principles used in motor vehicle systems. Additional material is included to give a more detailed understanding of the electrical components used in these systems, and the treatment is such that the book constitutes an ideal aid for students following a City and Guilds or BTEC course in Motor Vehicle Studies.

For simplicity, the vehicle electrical system has been divided into five sub-systems: lighting, ignition, starting, charging and ancillaries.

The book has been written on the assumption that the reader will possess a basic knowledge of mathematics.

Some worked examples are included in the text to illustrate the principles discussed and a number of exercises are provided at the end of the book. Answers to numerical problems are given.

Many thanks are necessary to my wife, Lesley, for her perseverance with the typing.

J H

1 Introduction to electricity

1.1 Some basic concepts

Sources of electricity, such as batteries and generators, are said to produce an **electromotive force** (e.m.f.) E, which is measured in **volts** (V). This voltage is the force (potential) which causes an electric current to flow.

Atoms contain minute particles called **protons**, which have a **positive charge**, and **electrons** which have a **negative charge**. Metals, such as copper, have many electrons which are held loosely by their atoms. These electrons move haphazardly from atom to atom and are known as **free electrons** (Fig. 1.1a).

One of the basic laws of magnetism is that unlike poles attract each other, i.e. north to south, and like poles repel each other, i.e. two north or two south poles. Similarly, in electricity, unlike charges are attracted to each other and like charges are repelled from each other. If an energy source, such as a battery, is connected to a metal wire, then the electromotive force acts on the free electrons in such a way that they are attracted towards the positive terminal and drift along the wire in one direction, i.e. from the negative terminal to the positive terminal. This movement of electrons is known as an electric current, and because it flows in one direction only, is referred to as **direct current** (see Fig. 1.1b).

Before the discovery of the electron, it was believed that electric current flow was due to positive electric charges moving round the circuit from the positive terminal to the

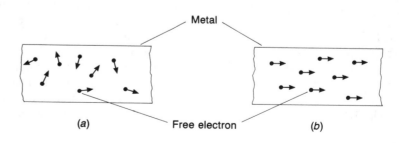

Fig. 1.1
(a)   Free electron   (b)

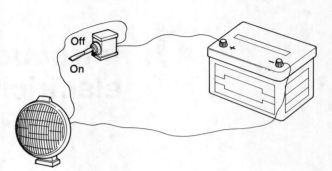

Fig. 1.2

negative terminal. This concept is maintained today and is known as **conventional current flow**, i.e. the direction in which positive charges would flow, and is normally indicated on circuit diagrams by an arrow.

The circuit shown in Fig. 1.2 consists of a battery, a switch and a lamp. When the switch is closed, the circuit has **continuity** and current will flow, as shown by the arrows, to light the lamp.

The unit for electric current I is the **ampere** (A), normally abbreviated to amp.

All electrical devices and components offer some resistance that limits the flow of current. The unit for this resistance R is the **ohm** (Ω). It is often necessary to add resistance to a circuit in order to limit the size of current flowing when a particular value of voltage is applied to the circuit. For this reason, components named **resistors** are manufactured which have a fixed value of resistance; alternatively, variable resistors are available whose resistance can be changed to give a specific value of resistance. The symbols used for these resistors in circuit diagrams are shown in Fig. 1.3.

The amount of voltage, current and resistance in a circuit can be measured using instruments known respectively as **voltmeters**, **ammeters** and **ohmmeters**. The manner in which a voltmeter and ammeter are connected is important and is illustrated in Fig. 1.4. An ammeter is connected in **series** with the battery terminals and the other components of the circuit whereas a voltmeter is connected in **parallel** or **across** the component(s) whose voltage is to be measured.

An ohmmeter is connected in the same way as a voltmeter, but it must be ensured that no current is flowing when resistance checks are made or damage to the instrument can occur.

These three instruments can be combined into one instrument called a **multimeter**. Multimeters are manufactured in two forms: **analogue** or **digital**. Analogue meters use a needle deflection for measurement, whereas digital meters give a

(a)   (b)

Fig. 1.3 (a) Fixed resistor, (b) Variable resistor

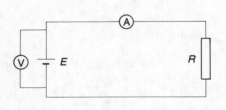

Fig. 1.4

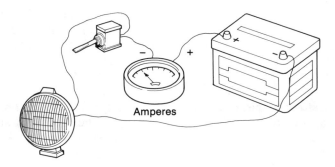

Fig. 1.5

numerical display. Analogue meters generally depend on current passing through them in a certain direction to ensure correct operation and their terminals are therefore marked + and − to show how they should be connected with respect to the terminals of the supply, as shown in Fig. 1.5.

1.2 The battery

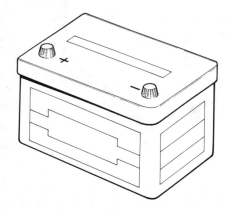

Fig. 1.6 The lead-acid vehicle battery

Batteries are usually made by connecting **cells** together, the cells used being divided into two principal groups: **primary cells** and **secondary cells**. The main difference between the two groups is that secondary cells are **discharged** when being used and can be **recharged** as necessary, whereas primary cells lose their e.m.f. when their energy is exhausted and cannot be recharged.

Batteries made from secondary cells are in turn divided into two main groups: **lead acid** and **alkaline**, and vehicle systems invariably employ the lead acid type. The vehicle battery is the main energy source for the vehicle electrical systems. It will take energy from the generator when the engine is running at normal speed and it will supply energy to the electrical systems when the engine is switched off. Figure 1.6 shows a typical vehicle battery. It has two terminals whose **polarity** is indicated by + (positive) and − (negative). Some battery terminals are painted red (positive) and blue (negative). The positive terminal is usually larger in size than the negative terminal in order to avoid any incorrect connections.

Vehicle electrical systems are designated as being either positive or negative **earth return**. This determines which terminal is connected directly to the vehicles metal chassis (earth) and care must be taken when connecting a battery that the correct terminal is earthed.

Batteries are referred to by their terminal voltage and vehicle batteries are usually manufactured to provide 6, 12 or 24 volts. Thus most cars have a 12 volt, lead-acid battery.

The symbols used in electrical circuits for a battery are shown in Fig. 1.7, where the longer lines represent the positive

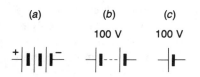

Fig. 1.7 Battery symbols

plates. Figure 1.7*a* represents a battery of 6 volts having three 2 V cells. When multi-cell batteries are employed the symbols shown in Fig. 1.7*b* or 1.7*c* are used with the e.m.f. written above, as illustrated.

1.3 Battery construction and operation

A typical construction of a lead acid battery is shown in Fig. 1.8. The term lead acid is used because the battery uses lead and sulphuric acid as active materials.

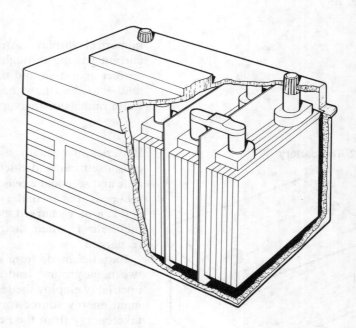

Fig. 1.8

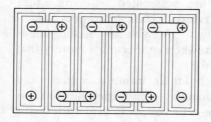

Fig. 1.9

The battery consists of numbers of separate cells, each with its own positive and negative terminals, connected together as shown in Fig. 1.9. This series connection of cells, i.e. the negative terminal of one linked to the positive terminal of the next, means that the terminal voltage of the battery is equal to the sum of the cell voltages. Each cell has an e.m.f. of approximately 2 V and Fig. 1.9 therefore represents a battery with a terminal voltage of 12 V. The cells are housed in an acid-proof casing.

Each individual cell has two sets of plates which are suspended in dilute sulphuric acid, known as the **electrolyte**. The plates connected to the positive terminal contain lead peroxide and those connected to the negative terminal contain lead in a spongy porous form. A chemical reaction occurs as a result of dissimilar metals being in contact and the acid and chemical energy is converted into electrical energy. Figure

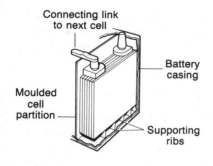

Fig. 1.10

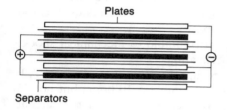

Fig. 1.11

1.10 shows the typical construction of a single lead-acid cell.

The negative plates experience a greater chemical reaction than the positive plates and it is usual to include one extra negative plate in each cell to take account of this effect.

Porous separators, which allow the electrolyte to circulate fully, ensure against internal short circuits between the plates.

A vent is also provided at the top of each cell to allow for gases to escape and for adjustment of the electrolyte.

Secondary batteries **discharge** electrical energy to loads connected to their terminals and have their energy **recharged** using a charging circuit. A process called **electrolysis** occurs when a battery is discharging or charging. Electrolysis is the name given to the chemical changes which occur when an electric current is passed through an electrolyte. Electrolysis in a lead-acid battery is effectively the separation of water into hydrogen and oxygen and during discharge, when current flows from the battery, the battery plates become coated with lead sulphate, the acid becomes diluted and hydrogen gas is given off. These electrochemical reactions are reversed during the recharging process when current flows through the battery from an external source in that the lead sulphate is removed from the plates and returned to the acid, with hydrogen again being given off. The battery can therefore be returned to its original state.

If a fully serviceable battery is left unused it will self discharge over a long period of time (usually weeks), due to various reasons such as current leakage between cells. Unless the battery is recharged, **sulphation** occurs in which lead sulphate forms on the plates in the form of a hard white crust which increases the internal resistance of the battery to a maximum and this in turn prevents any current from flowing through the battery. Unless the lead sulphate is removed by charging, the battery will become unserviceable.

The **capacity** for a battery to store energy is measured in **ampere-hours** and is determined by discharging a fully charged battery at a constant current rate through a load until each cell voltage falls to approximately 1.8 V. The capacity is then calculated by multiplying the current in amperes by the time in

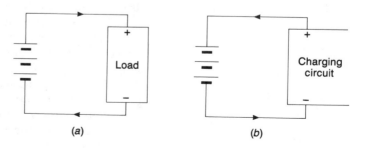

Fig. 1.12 (a) Discharge and (b) charge circuits

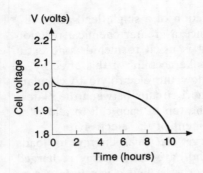

Fig. 1.13 Typical cell discharge characteristic

hours. A general discharge time is 10 hours and Fig. 1.13 illustrates typical cell discharge characteristics, showing the variation of voltage with time when being discharged at a constant rate.

The capacity is dependent on various factors.

1 Acid strength. If this is too strong the plates are damaged, and if too weak the performance is reduced due to the limitations on the chemical process.
2 Cold temperatures reduce battery capacity and either stronger acid is required or some means of warming the battery.
3 Sulphation and low electrolyte levels effectively reduce the surface area of the plates and, as capacity is approximately proportional to the total surface area of the plates, this too will reduce the battery capacity.
4 Reduction in capacity is attributable to mechanical shock, rough treatment and overheating.
5 Discharge rate also affects capacity. For example, a 35 A h battery may well provide 3.5 amps for 10 hours but not 35 amps for 1 hour, due to overheating and the large current flow causing plate disintegration.

1.4 Battery charging

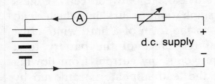

Fig. 1.14

A basic charging circuit as shown in Fig. 1.14, consisting of a 6 V battery connected in series to a d.c. supply via an ammeter and a variable resistor. The positive connection of the supply is connected to the battery positive terminal and the negative to the negative terminal. The voltage available to provide charging current is therefore the difference between the supply voltage and the battery voltage and consequently a larger supply voltage than battery voltage is required for current to flow in the direction shown.

1.5 Potential difference

The variable resistor in Fig. 1.14 controls the charging current by controlling the electrical pressure across the battery terminals. If current flows through a resistor then a **voltage drop** occurs across the resistor and a **potential difference** (p.d.) is said to exist across it.

Potential difference is measured in volts, as is e.m.f. However, whereas e.m.f. refers to the voltage available at the supply voltage of a source, p.d. refers to the voltage(s) dropped across component(s) in a circuit. This is illustrated in Fig. 1.15 where voltmeter V_1, is measuring terminal voltage and voltmeters V_2 and V_3 are measuring the potential differences across the resistors. The letter symbols for e.m.f. and p.d. are E and V respectively.

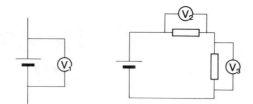

Fig. 1.15

1.6 Specific gravity and the hydrometer

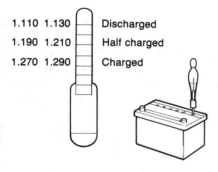

Fig. 1.16

Batteries are checked periodically during charging by using a **hydrometer** to measure the **specific gravity** of the electrolyte. As acid strength is increasing during charging, the specific gravity is also increasing.

Hydrometers are either graduated with typical values, as shown in Fig. 1.16, for various states of battery charge or, alternatively, they may have a colour sequence with green indicating full charge, yellow indicating half charge and red indicating a flat battery. The values of specific gravity shown in Fig. 1.16 refer to a temperature of 15 °C, since specific gravity varies with temperature.

2 Resistors and Ohm's law

2.1 Construction of resistors

Resistors are manufactured in various ways. Some consist of special resistance wire wound on to an insulated former and others are made largely of carbon. Other types are made by depositing carbon or a film of metal on to a porcelain tube. Figure 2.1 shows some examples.

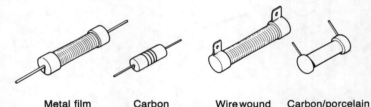

Fig. 2.1

Metal film   Carbon   Wire wound   Carbon/porcelain

The resistance value of an individual resistor is sometimes printed on the component. However, this is not always practicable as components may be small or awkwardly mounted, or indeed discoloured by heat. Most resistors, therefore, use a **colour code** denoting their resistance value.

In addition to the resistance value, the colour code also identifies the **tolerance** of the resistor. Because of the difficulty in manufacturing resistors to an exact value, they are constructed to various tolerances. These tolerances are 20%, 10%, 5%, 2%, and 1%. Therefore, a 100 ohm resistor with a 10% tolerance has an actual value lying between 90 ohms and 110 ohms.

2.2 Colour coding of resistors

The general format for the British colour coding of resistors is to use four colours. Three are used for the resistance value and the fourth is used to denote the tolerance. Two methods of colour coding are shown in Fig. 2.2. Of all methods, the band method is the most commonly used. The resistance colour code is shown in Table 2.1.

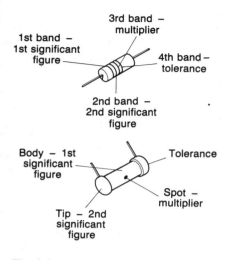

Fig. 2.2

Table 2.1 Resistor colour codes

Colour	Band 1 1st digit	Band 2 2nd digit	Band 3 multiplying factor	Band 4 tolerance
Black	–	0	1	–
Brown	1	1	10	1%
Red	2	2	100	2%
Orange	3	3	1000	–
Yellow	4	4	10 000	–
Green	5	5	100 000	–
Blue	6	6	1 000 000	–
Violet	7	7	–	–
Grey	8	8	–	–
White	9	9	–	–
Gold	–	–	0.1	5%
Silver	–	–	0.01	10%
No colour	–	–	–	20%

Example 2.1

A resistor has four bands coloured blue, grey, red and silver. The resistance value of the component is:

Blue	Grey	Red	Silver
6	8	×100	±10%

i.e. 6800 ohms ± 10%

To avoid the manufacture of many different values of resistor, carbon resistors are usually made in a range of **preferred values**, which use the tolerance factors to ensure that all values are covered. For example, the 10% series has values of 10, 12, 15, 18, 22, 27, 33, 39, 47, 56, 68 and 82 ohms, repeating in multiples of 10, i.e. 100, 120, 150, etc.

If, therefore, a 36 ohm resistor is required, then a 33 or 39 preferred value can be used:

33 + 10% = 36.3 ohms
39 − 10% = 35.1 ohms

When selecting a specific resistor for use in a particular circuit, its power rating must be considered in addition to its resistance value. This power rating is the maximum power that the resistor can dissipate safely.

2.3 Prefixes

Considerable use is made of prefixes to represent decimal multiples or sub-multiples of electrical and electronic units. The most commonly used prefixes and their values are shown in Table 2.2.

Table 2.2 Prefixes for electrical units

Value	Unit	Symbol
Multiples		
$\times 10^6$	mega	M
$\times 10^3$	kilo	k
Sub-multiples		
$\times 10^{-3}$	milli	m
$\times 10^{-6}$	micro	μ
$\times 10^{-9}$	nano	n
$\times 10^{-12}$	pico	p

2.4 British Standard resistance code

Table 2.3

Resistance value (Ω)	Code
0.33	R33
1	1R0
3.3	3R3
33	33R
100	100R
1k	1K0
10k	10K
10M	10M

2.5 Ohm's law

Resistance values are identified by using BS1852. This is an alphanumeric code and its application is shown in Table 2.3.

Tolerances are also identified by letters:

F = ±1% G = ±2% J = ±5% K = ±10% M = ±20%

Examples are:

R33M = 0.33 Ω ±20%
3R7K = 3.7 Ω ±10%
5K6F = 5.6 k Ω ±1%

The relationship between an applied e.m.f. and the current flow in a conductor was investigated by a Professor S. G. Ohm. The result of his experiments proved that for a given conductor at a constant temperature, the ratio

$$\frac{\text{applied e.m.f. in volts}}{\text{current flow in amperes}}$$

was constant for a range of applied voltages. This ratio is known as the resistance of the conductor to current flow and is measured in ohms (Ω).

Ohm's law may be stated as, 'In a conductor at uniform temperature, the current flow is directly proportional to the potential difference between its ends, providing that the resistance in the circuit remains constant.'

Mathematically, Ohm's law can be abbreviated to:

$$\frac{\text{volts}}{\text{amperes}} = \text{ohms}$$

By transformation of formulae, it can also be expressed as:

volts = amperes × ohms

or

$$\text{amperes} = \frac{\text{volts}}{\text{ohms}}$$

A simple memory aid is shown in Fig. 2.3. From this, to find the value of I, simply cover it with a finger and the expression is obtained from the two remaining symbols:

$$I = \frac{E}{R}$$

The same principle is adopted to find the expressions for E and R.

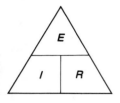

Fig. 2.3

2.6 Simple circuit analysis

Example 2.2

When an e.m.f. of 6 V is applied to a circuit, the current flow is measured as 0.5 A. What is the resistance of the load?
From the expression

$$R = \frac{E}{I}$$

$$R = \frac{6}{0.5}$$

$$R = 12 \text{ ohms.}$$

The resistance of the load is therefore 12 ohms.

Example 2.3

A voltage of 250 V is applied to a circuit which has a resistive load of 1000 Ω. What current will flow through the load?
From the expression

$$I = \frac{E}{R}$$

$$I = \frac{250}{1000}$$

$$I = 0.25 \text{ amps.}$$

The current through the load will be 0.25 amps.

3 Resistors in series

3.1 Introduction

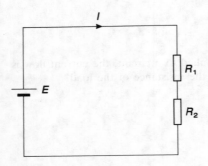

Fig. 3.1

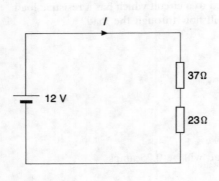

Fig. 3.2

Resistors are used to limit the size of current flow in an electrical circuit. However, it is sometimes not possible, or indeed practical, simply to use one resistor to achieve the desired result and it becomes necessary to employ two or more resistors.

Figure 3.1 shows a circuit which has a **load** made up of two separate resistors. These resistors are said to be connected in **series** with the supply.

Note: The meaning of load is important and is often misunderstood. Load implies work and the work done in a resistive circuit is the movement of electrons which comprise the current. The load of a circuit is actually the current. A large load means that a large current flows in the circuit, i.e. low circuit resistance, whereas a light load coincides with a high circuit resistance.

The current in the circuit of Fig. 3.1 has only one path in which to flow and therefore the same current passes through both resistors. Ohm's law states that:

$$I = \frac{E}{R}$$

where R is the total resistance of the circuit. The total resistance R_T for a number of series-connected resistors is equal to the sum of the individual resistances. For Fig. 3.1

$$R_T = R_1 + R_2$$

Example 3.1

Consider the circuit in Fig. 3.2. An e.m.f. of 12 volts is applied to the ends of two resistors connected in series. The values of the resistors are 37Ω and 23Ω. What current will flow through the circuit?

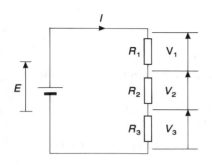

Fig. 3.3

The total resistance

$$R_T = 37 + 23 \text{ ohms}$$

$$R_T = 60 \text{ ohms}$$

Using Ohm's law, the current I through the circuit will be

$$I = \frac{E}{R}$$

$$I = \frac{12}{60}$$

$$I = 0.2 \text{ amps.}$$

3.2 Voltages in a series circuit

Figure 3.3 shows a circuit with three resistors connected in series. Because current flows through each resistor, a p.d. will exist across each one. These p.d.s are found by applying Ohm's law.

$$V_1 = I \times R_1$$

$$V_2 = I \times R_2$$

and

$$V_3 = I \times R_3$$

A fundamental law in electricity is **Kirchhoff's voltage law** and this states that the algebraic sum of the voltages in any closed circuit is zero. Therefore, for Fig. 3.3,

$$E = V_1 + V_2 + V_3.$$

Substituting for V_1, V_2 and V_3 gives

$$E = IR_1 + IR_2 + IR_3$$

$$E = I(R_1 + R_2 + R_3).$$

Example 3.2

With reference to Fig. 3.3, when the supply voltage is 24 V, $R_1 = 12\Omega$, $R_2 = 7\Omega$ and $R_3 = 11\Omega$, determine the current flowing in the circuit and the p.d. across each resistor.

Total resistance

$$R_T = R_1 + R_2 + R_3$$

$$R_T = 12 + 7 + 11 \text{ ohms}$$

$$R_T = 30 \text{ ohms.}$$

The circuit current

$$I = \frac{E}{R}$$

$$I = \frac{24}{30}$$

$I = 0.8$ amps.

This current of 0.8 A flows through each resistor. Therefore:

The p.d. across $R_1 = 0.8 \times 12 = 9.6$ V;
The p.d. across $R_2 = 0.8 \times 7 = 5.6$ V; and
The p.d. across $R_3 = 0.8 \times 11 = 8.8$ V.

These values can be verified by applying Kirchhoff's law, i.e.

$$E = V_1 + V_2 + V_3$$

$$E = 9.6 + 5.6 + 8.8$$

$$E = 24 \text{ volts.}$$

An alternative method of calculating p.d.s across the resistors without first calculating the current can be used if the supply voltage is known, using the **potential divider rule**. It has been shown for Fig. 3.3 that

$$E = IR_1 + IR_2 + IR_3,$$

and

$$I = \frac{E}{R_T}$$

Substituting for I in the expression for E gives

$$E = \left(\frac{E}{R_T} \right) R_1 + \left(\frac{E}{R_T} \right) R_2 + \left(\frac{E}{R_T} \right) R_3$$

or

$$E = E \left(\frac{R_1}{R_T} \right) + E \left(\frac{R_2}{R_T} \right) + E \left(\frac{R_3}{R_T} \right)$$

but

$$E = V_1 + V_2 + V_3,$$

hence

$$V_1 = E \left(\frac{R_1}{R_T} \right), \ V_2 = E \left(\frac{R_2}{R_T} \right), \ V_3 = E \left(\frac{R_3}{R_T} \right)$$

giving

$$V_1 = 24 \times \frac{12}{30} = 9.6 \text{ V}$$

$$V_2 = 24 \times \frac{7}{30} = 5.6 \text{ V}$$

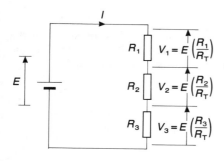

Fig. 3.4

$$V_3 = 24 \times \frac{11}{30} = 8.8 \text{ V}$$

as obtained previously. These voltages are illustrated in Fig. 3.4.

Example 3.3

Three resistors are connected in series across a 48 V supply. Their values are 9Ω, 12Ω and 15Ω. Determine the p.d. across each resistor using the potential divider rule.

$$R_T = 9 + 12 + 15 = 36 \text{ ohm}$$

$$\text{p.d. across the } 9 \text{ } \Omega \text{ resistor} = 48 \times \frac{9}{36}$$

$$= 12 \text{ V}$$

$$\text{p.d. across the } 12 \text{ } \Omega \text{ resistor} = 48 \times \frac{12}{36}$$

$$= 16 \text{ V}$$

$$\text{p.d. across the } 15 \text{ } \Omega \text{ resistor} = 48 \times \frac{15}{36}$$

$$= 20 \text{ V.}$$

3.3 Internal resistance

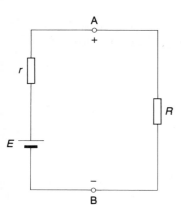

Fig. 3.5

All batteries have some internal resistance. Where this resistance is small in comparison with the circuit resistance, its effects can usually be ignored, but if consideration is to be given to its effects then it can be represented as a fixed resistor in a circuit diagram as shown in Fig. 3.5.

The points A and B represent the battery terminals with the circuit drawn to the left of these points representing the e.m.f. of the battery in series with its internal resistance r. A p.d. will be dropped across both resistors (V_r and V_R).

Therefore,

$$E = V_r + V_R \text{ (Kirchhoff's voltage law)}.$$

If the e.m.f. is constant, then current I will be constant and

$$E = Ir + IR$$

$$E = I(r + R)$$

The current in the circuit

$$I = \frac{E}{r + R} \text{ amps.}$$

Example 3.4

An electrical supply has an e.m.f. of 48 V. The resistance of the circuit connected to the supply terminals is 220Ω and the internal resistance is 20Ω. What current flows in the circuit?

$$I = \frac{E}{r + R} \; .$$

Therefore

$$I = \frac{48}{20 + 220}$$

$$I = \frac{48}{240}$$

$$I = 0.2 \text{ A.}$$

3.4 Battery terminal voltage

From the earlier expression $E = V_r + V_R$

$$V_R = E - V_r.$$

The significance of this is that if the internal resistance V_r is small compared with V_R then it can be ignored and, practically, $V_R = E$.

If, however, the resistance of the external circuit decreases then more current flows, the p.d. across V_r increases and the terminal p.d. decreases. This is illustrated in Example 3.5.

Example 3.5

An electrical supply of 40 V has an internal resistance of 100Ω. What is the terminal voltage when the load resistance is:

a) 7k9Ω?
b) 700Ω?

With a load of 7k9Ω

$$I = \frac{E}{r + R}$$

$$= \frac{40}{100 + 7900}$$

$$I = 0.005 \text{ A} = 5 \text{ mA.}$$

$$\text{Terminal voltage} = 5 \text{ mA} \times 7k9Ω$$

$$= 39.5 \text{ V.}$$

With a load of 700Ω

$$I = \frac{E}{r + R}$$

$$= \frac{40}{100 + 700}$$

$$I = 0.05 \text{ A} = 50 \text{ mA}.$$

Terminal voltage $= 50 \text{ mA} \times 700\Omega$

$$= 35 \text{ V}.$$

If the internal resistance of the supply is high in comparison with the load resistance, then quite large changes in the load will only produce small changes in load current. In practice, the internal resistance of batteries and generators is kept very small (fractions of an ohm) to prevent the source from overheating and increasing the efficiency of the supply.

Example 3.6

A d.c. generator having a terminal voltage of 250 V has an internal resistance of 2k2Ω. What current will flow through a load of:

a) 300Ω?
b) 700Ω?

With a load of 300Ω

$$I = \frac{250}{2200 + 300}$$

$$= 0.1 \text{ A}.$$

With a load of 700Ω

$$I = \frac{250}{2200 + 700}$$

$$= 0.086 \text{ A}.$$

d.c. supplies normally have low internal resistance and thus act as constant voltage generators. Example 3.6 illustrates a source which can be considered to be a constant current supply, but these occur rarely in practice.

4 Resistors in parallel

4.1 Introduction

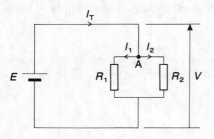

Fig. 4.1

Just as adding resistors in series increases the resistance of a circuit, the resistance of a circuit can be reduced by connecting the resistors in **parallel** with the supply. Figure 4.1 shows a circuit with two resistors connected in this way. The p.d.s across each resistor will be equal to each other and will also be equal to the supply e.m.f.

Kirchhoff's current law states that the sum of the currents leaving a junction equals the sum of the currents entering the junction. From Fig. 4.1, at point A

$$I_T = I_1 + I_2 \tag{4.1}$$

As $E = V$, (by Kirchhoff's voltage law), and $I = E/R$ (Ohm's law) then

$$I_T = \frac{E}{R_T} \, ; \, I_1 = \frac{E}{R_1} \, ; \, I_2 = \frac{E}{R_2} \, ,$$

where R_T is the effective circuit resistance. Substituting in equation (4.1)

$$\frac{E}{R_T} = \frac{E}{R_1} + \frac{E}{R_2} \, .$$

Dividing the equation throughout by E

$$\frac{1}{R_T} = \frac{1}{R_1} + \frac{1}{R_2} \, .$$

Example 4.1

A circuit consists of three resistors connected in parallel. The values of the resistors are 6Ω, 9Ω and 18Ω. If 15 V is applied to the circuit, determine:

a) the total circuit resistance,
b) the current flowing through each resistor, and
c) the circuit current.

a)
$$\frac{1}{R_T} = \frac{1}{R_1} + \frac{1}{R_2} + \frac{1}{R_3}$$

$$= \frac{1}{6} + \frac{1}{9} + \frac{1}{18}$$

$$= \frac{3 + 2 + 1}{18}$$

$$= \frac{6}{18} .$$

$$R_T = \frac{18}{6} = 3\Omega.$$

b)
$$I_1 = \frac{E}{R_1} = \frac{15}{6} = 2.5 \text{ A}.$$

$$I_2 = \frac{E}{R_2} = \frac{15}{9} = 1.67 \text{ A}.$$

$$I_3 = \frac{E}{R_3} = \frac{15}{18} = 0.83 \text{ A}.$$

c)
$$I_T = I_1 + I_2 + I_3$$

$$= 2.5 + 1.67 + 0.83$$

$$= 5 \text{ A}.$$

4.2 The product over sum rule

When there are only two resistors connected in parallel, the total circuit resistance can be calculated by using the product over sum rule. Assume that two resistors R_1 and R_2 are connected in parallel. Then

$$\frac{1}{R_T} = \frac{1}{R_1} + \frac{1}{R_2}$$

$$\frac{1}{R_T} = \frac{R_1 + R_2}{R_1 \times R_2} .$$

Therefore

$$R_T = \frac{R_1 \times R_2}{R_1 + R_2} .$$

This rule can be used where there are more than two resistors.

Example 4.2

Three resistors, whose values are 100Ω, 150Ω and 180Ω, are connected in parallel. What is the total resistance of the network?

Consider first the 100Ω and the 150Ω resistors. Let their equivalent resistance be R_p

$$R_p = \frac{100 \times 150}{100 + 150} = 60\Omega.$$

The network can now be considered as a 60Ω resistor in parallel with a 180Ω resistor. Therefore

$$R_T = \frac{60 \times 180}{60 + 180} = 45\Omega.$$

4.3 Resistors of the same value

If a number of resistors of equal value are connected in parallel, then

$$\text{Total circuit resistance} = \frac{\text{Value of 1 resistor}}{\text{Number of resistors}}.$$

Example 4.3

When five 20Ω resistors are connected in parallel, what is the total resistance?

$$R_T = \frac{20}{5}$$

$$R_T = 4\Omega.$$

This can be proved by using the basic formula given in Section 4.1

$$\frac{1}{R_T} = \frac{1}{20} + \frac{1}{20} + \frac{1}{20} + \frac{1}{20} + \frac{1}{20}$$

$$\frac{1}{R_T} = \frac{5}{20}$$

$$R_T = \frac{20}{5}$$

$$R_T = 4\Omega.$$

4.4 Meter resistance

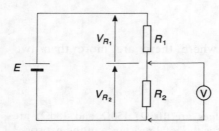

Fig. 4.2

Figure 4.2 shows a voltmeter connected to measure the voltage dropped across resistor R_2. When an analogue multimeter is used for this purpose, then it will only operate when current passes through it. The voltmeter itself possesses resistance, and this resistance is now effectively connected in parallel with R_2. The effect of this is shown in Example 4.4.

Example 4.4

Referring to Fig. 4.2, the applied e.m.f. is 30 V, $R_1 = 10\text{k}\Omega$ and $R_2 = 5\text{k}\Omega$. The p.d. across R_2 can be determined using the potential divider rule.

$$V_{R_2} = E \times \frac{R_2}{R_1 + R_2}$$

$$= 30 \times \frac{5}{10 + 5}$$

$$V_{R_2} = 10 \text{ V}.$$

Assume now that an analogue multimeter is connected across R_2 to measure the voltage drop and that the internal resistance of the meter is 50kΩ. The effective resistance R_e of the meter and R_2 in parallel is

$$R_e = \frac{5 \times 50}{5 + 50}$$

$$R_e = 4.55\text{k}\Omega.$$

Using the potential divider rule

$$V_{R_2} = 30 \times \frac{4.55}{10 + 4.55}$$

$$V_{R_2} = 9.4 \text{ V}.$$

The meter resistance acts as a **shunt** on the circuit and can result in inaccurate readings. Modern-day multimeters have a much higher internal resistance than the 50kΩ used in the above example and this figure has been used purely to illustrate the shunting effect.

4.5 Resistors in series/parallel combinations

To calculate the total circuit resistance when resistors are connected in both series and parallel, it is necessary to reduce each parallel network to a single resistance value and then to treat the complete circuit as for resistors connected in series.

Example 4.5

Calculate the total resistance of the network shown in Fig. 4.3. Let the effective resistance of the parallel resistors be R_e.

$$R_e = \frac{48 \times 12}{48 + 12}$$

$$= \frac{576}{60}$$

$$R_e = 9.6\text{k}\Omega$$

The total resistance

$$R_T = 9.6 + 14$$

$$R_T = 23.6\text{k}\Omega.$$

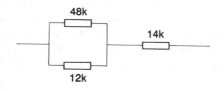

Fig. 4.3

5 Electrical power

5.1 Electrical quantity

Chapter 1.3 refers to the ampere-hour as a measure for a battery's capacity. The ampere-hour is a means of indicating how much energy is available for a specified period of time when a circuit is connected to the battery terminals. It is a basic unit of electrical quantity, the international unit being the **coulomb**. One coulomb of charge is said to flow if a wire carries one ampere of current for one second. Energy is defined as the capacity for doing work or, similarly, work is done when energy is transferred.

5.2 Power rating

The rate of doing work in electrical applications is known as the **power rating** of the load. A vehicle headlamp provides illumination for driving at night and in conditions when visibility is poor. Electrical energy from the battery is converted by the bulb filament into light and heat. Panel lights inside the vehicle operate in the same manner but to a less powerful degree. These components are given a power rating which provides an indication of their relative outputs compared to each other. For a headlamp, this rating may typically be in the order of 60 watts on main beam and 50 watts when dipped. The **watt** (W) is the basic unit of **power** P and is named after James Watt who invented the steam engine. Obviously, the higher the power rating, the larger the amount of energy transferred and subsequently the more work done. The three factors, power, work (energy transferred) and time are interrelated. Mathematically

$$\text{power} = \frac{\text{work}}{\text{time}},$$

where power is measured in watts (W), work is measured in joules (J), and time is measured in seconds (s).

5.3 The power formulae

The coulomb is expressed mathematically as $Q = It$ coulombs. If one joule of energy is consumed when one coulomb flows between two points, then a p.d. of one volt exists between those points.

$$\text{Voltage } (V) = \frac{\text{work } (w)}{\text{quantity } (Q)}$$

or

$$w = QV.$$

Substitute $Q = It$

$$w = ItV$$

or

$$\frac{w}{t} = IV.$$

Therefore,

$$\text{Power } (P) = I \times V \text{ watts}$$

Example 5.1

A headlamp bulb rated at 36 W is supplied with 12 V from the battery. What current will flow when the headlamp is switched on?

$$P = I \times V$$

$$I = \frac{P}{V}$$

$$I = \frac{36}{12} = 3 \text{ A.}$$

When using a battery supply it is useful to know for how long a battery can continue to supply current to a load.

Example 5.2

A battery maintains a constant voltage of 12 V when supplying a headlamp rated at 36 W. If the battery has a capacity of 200 A h, how long will it take to reduce the charge on the battery to one-half of full charge?

$$\text{Half charge} = \frac{200}{2} = 100 \text{ A h.}$$

$$I = \frac{P}{V} = \frac{36}{12} = 3 \text{ A.}$$

$$\text{Time} = \frac{100 \text{ A h}}{3 \text{ A}} = 33.3 \text{ hours.}$$

5.4 Alternative forms of the power formula

$$P = V \times I \tag{5.1}$$

From Ohm's law

$$V = I \times R \tag{5.2}$$

Substitute the expression for V from (5.2) in (5.1)

$$P = I \times R \times I$$

$$P = I^2 R$$

Similarly, from Ohm's law,

$$I = \frac{V}{R}$$

Substituting this expression for I in equation (5.1)

$$P = V \times \frac{V}{R}$$

$$P = \frac{V^2}{R} .$$

Example 5.3

In a 12 V vehicle system a headlamp has a resistance of 4Ω. What is the power rating of the headlamp?

$$P = \frac{V^2}{R}$$

$$= \frac{12 \times 12}{4}$$

$$= \frac{144}{4}$$

$$P = 36 \text{ W.}$$

6 Magnetism and electromagnetism

6.1 Introduction

Some of the effects of magnetism have been known to man for more than 2000 years. Permanent magnets have been used as compass needles for navigational purposes for hundreds of years due to their interaction with the earth's magnetic field. The ability to produce electricity from magnetism was discovered by Michael Faraday in the 19th century when he invented the generator. Some of the properties of magnets employed in vehicle electrical systems are:

a) Magnets exert a force of attraction on materials such as iron and steel.
b) Like magnetic poles (north–north, south–south) are repelled from each other.
c) Unlike magnetic poles (north–south) produce a force of attraction to each other.

Magnetic lines of force are considered to act in a direction from a north pole to a south pole.

6.2 Magnetism and electric current

When a current flows through a conductor, a magnetic field is set up around the conductor in the form of a series of concentric circles. The density of this field decreases as the distance from the conductor increases. The **corkscrew rule** is used to determine the direction in which the field acts and this depends on the direction of current flow.

If a corkscrew is turned so that it moves through the conductor in the same direction as the current, then the direction in which it is turned is the direction in which the magnetic field acts.

When a current-carrying conductor is formed into a loop or single coil, the magnetic lines of force are concentrated in a circular space with their **flux density** being greatest in the centre of the coil. Figure 6.2 represents a single coil in cross-section. The cross symbol at the bottom indicates current flow

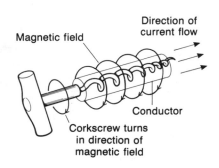

Magnetic field

Direction of current flow

Conductor

Corkscrew turns in direction of magnetic field

Fig. 6.1

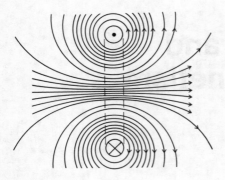

Fig. 6.2

into the wire and the dot symbol at the top indicates current flow out of the wire. These two points can be considered as two separate, straight, current-carrying conductors, and by applying the corkscrew rule to both points it is apparent that the magnetic fluxes produced will be additive and act in the same direction in the centre of the coil.

By winding a conductor into a number of turns, a **solenoid** is produced. If the turns are wound closely, or in several layers, the flux passes through the centre of the solenoid, linking with all of the turns. As with a single coil, the flux density is greatest in the centre of the solenoid. Its direction is determined by a grip rule which states that if the four fingers are held in the direction of conventional current flow, then the thumb indicates the direction of flux inside the solenoid (Fig. 6.3).

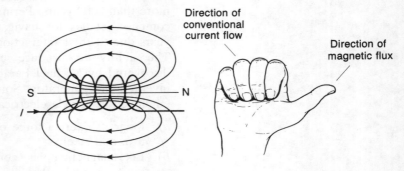

Fig. 6.3

The flux density inside a solenoid, or **electromagnet**, can be increased by placing a soft iron bar in its centre or by increasing the current and/or the number of turns.

6.3 The electrical relay

A relay, as shown in Fig. 6.4, uses a solenoid for its operating principles. When an energising current is applied to the solenoid, the armature moves towards it due to magnetic attraction. This armature is pivoted such that its movement causes the contacts to close to complete an electrical circuit.

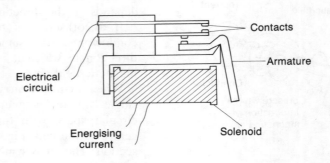

Fig. 6.4

The contacts remain closed until the solenoid is de-energised. Relays can have more than one contact and have many uses.

6.4 Electromagnetic induction

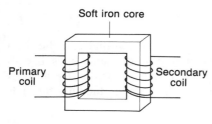

Soft iron core

Primary coil

Secondary coil

Fig. 6.5

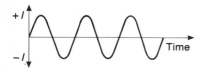

$+I$

$-I$

Time

Fig. 6.6

6.5 Self induction

The term electromagnetic induction is defined as the production of an e.m.f. in a conductor due to a changing magnetic field. One way in which electromagnetic induction can occur is by **flux linking** and this is demonstrated in the **transformer** (Fig. 6.4).

Two coils are wound on a soft iron core which has a magnetic flux established in it whenever current flows in either coil. If the primary coil is connected to a source of **alternating current** (a.c.), that is one which is continually changing in amplitude (Fig. 6.6), then the magnetic flux produced by this current is also changing continually. This changing flux passes through the core to the centre of the secondary coil (flux linking) and an e.m.f. is induced into the secondary coil. If this coil is connected to an external circuit then an alternating current will flow. Transformers can be used either to **step up** or to **step down** the input voltage by varying the number of primary, or more commonly, secondary turns. The output voltage V_s can be determined by the formula

$$V_s = V_p \times \frac{N_s}{N_p},$$

where V_p is the primary voltage, N_p is the number of primary turns, and N_s is the number of secondary turns.

The e.m.f. induced into the secondary coil of a transformer causes a current to flow in the secondary circuit. This current rises from zero and, in doing so, produces a changing magnetic field. The effect of this is to induce an e.m.f. back into the primary coil which in turn tries to set up another current in opposition to the initial primary current. This opposition is known as **reactance** and is summarised in Lenz's law, which states that the direction of an induced e.m.f. is such as to set up currents which cause a force opposing the change responsible for inducing the e.m.f. The unit of self inductance, or simply **inductance** L, is the **henry** and is given the symbol H. Inductance is present in all electrical circuits but is generally so small that it can be disregarded. In practice, the henry is a large unit of inductance and the millihenry (mH) and microhenry (μH) are more commonly used values. When conductors are arranged primarily to produce inductance, they are known as **inductors**. The amount of reactance X_L offered by inductors is measured in ohms and is frequency related.

$$X_L = 2\pi f L \text{ ohms,}$$

where f is the frequency in hertz, and L is the inductance in henrys.

At d.c. or low a.c. frequencies, the reactance is zero, or very low, whereas at high frequencies the reactance is large. It is generally accepted that inductors partially **block** a.c. but allow d.c. to flow easily.

6.6 Inductors connected in series and parallel

Inductors can be connected in series or parallel and the total inductance in the circuit is calculated on the same lines as with resistors similarly connected. For two inductors L_1 and L_2 connected in series or parallel, the total inductance (L_T) is given by

Series: $L_T = L_1 + L_2$ henrys

Parallel: $\dfrac{1}{L_T} = \dfrac{1}{L_1} + \dfrac{1}{L_2}$ henrys.

7 Capacitance

7.1 The capacitor

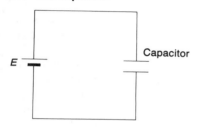

Fig. 7.1

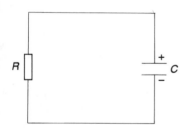

Fig. 7.2

A capacitor consists of two metal plates separated by an insulator known as a **dielectric**. Typical dielectric materials are mica, paper and even dry air. A capacitor has the ability to store an electric charge when it is connected to an e.m.f. Referring to Fig. 7.1, electrons will be attracted from the top plate of the capacitor to the positive terminal of the battery and through the circuit to the bottom capacitor plate. This leaves the top plate deficient in electrons (positively charged) and the lower plate with a surplus of electrons (negatively charged). This process is known as **charging**.

The flow of electrons continues until the potential difference between the capacitor plates is equal to E, when there is no longer any net e.m.f. round the loop to produce current. No actual current passes across the plates of the capacitor because of the dielectric between them.

If the battery is now removed, the charge remains on the plates and there will still be a potential difference between them. If a resistor is connected across the capacitor, as shown in Fig. 7.2, electrons flow from the negative plate back to the positive one until the p.d. (and hence the charge) is reduced to zero. The capacitor is now said to be discharging.

Figure 7.3a shows a graph of the rising voltage across the capacitor against time and Fig. 7.3b that of voltage against time for a discharging capacitor.

Fig. 7.3

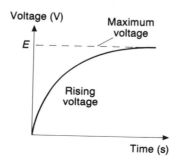
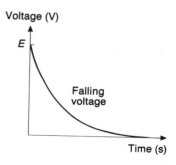

When direct voltages are applied to a capacitor, more charge is stored for a larger applied voltage. If a quantity of charge is given the symbol Q, then

$$Q = C \times V_C,$$

where C is a constant depending on the dimensions of the capacitor. If Q is in coulombs and V_C in volts, C will be in **farads** (F), and is called the **capacitance** of the capacitor.

The relationship between accumulated charge and current is given by

$$Q = I \times t,$$

where I is the current in amperes flowing into the capacitor and t is the time in seconds for which the current flows.

7.2 Capacitor charge

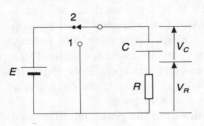

Fig. 7.4

All electric circuits contain resistance and the effects of charging a capacitor in a circuit which has resistance is illustrated in Fig. 7.4.

With the switch in position 1, any charge on C will leak away through R until C is fully discharged. Current ceases to flow and V_C and V_R will be zero. At the instant the switch is moved to position 2, V_C will be at zero as the capacitor has not had time to charge. The whole of the e.m.f. (E) will be dropped across the resistor and

$$V_R = E.$$

The current flow in the circuit is,

$$I = \frac{E}{R}.$$

Since $Q = I \times t$, the charge on C will increase as current flows and V_C will rise. This voltage opposes the battery voltage and

$$I = \frac{E - V_C - V_R}{R}.$$

As V_C continues to rise, then the current reduces and eventually when $V_C = E$, it falls to zero. V_R also falls to zero.

7.3 Time constant

The time constant used in electrical calculations and circuit design work for a capacitor connected in series with a resistor is determined by the formula

$$T = CR,$$

where T is the time constant in seconds, C is the capacitance

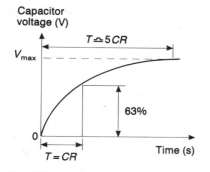

Capacitor voltage (V)

$T \simeq 5CR$

V_{max}

63%

$T = CR$

Time (s)

Fig. 7.5

of the capacitor in farads and R is the resistance of the resistor in ohms.

In actual fact, this time constant is the time taken for the voltage across the capacitor to rise to 0.632 of its final value (approximately 63%). The actual time taken to charge a capacitor is approximately $5CR$.

Example 7.1

A 0.5 µF capacitor is charged through a 2 MΩ resistor. What is the time constant of the circuit?

$$T = CR$$
$$= 0.5 \times 10^{-6} \times 2 \times 10^{6}$$
$$T = 1 \text{ s.}$$

7.4 Capacitors in series and parallel

It is sometimes necessary for capacitors to be combined to produce a larger or smaller capacitance. If two capacitors, C_1 and C_2, are connected in parallel, their capacitance values are added to give the total capacitance, C_T. Thus

$$C_T = C_1 + C_2 \text{ farads.}$$

If the same two capacitors are connected in series, the reciprocals of their capacitance values are added. The formula to calculate the total capacitance in this case is

$$\frac{1}{C_T} = \frac{1}{C_1} + \frac{1}{C_2} \text{ farads.}$$

7.5 Types of capacitor

Capacitors can be categorised as those which are **polarised**, and those which are not. A polarised capacitor is one which must be connected into a circuit in such a way that a specific plate of the capacitor is always positive with respect to the other plate. If this plate is not kept positive, the capacitor dielectric is destroyed and it will behave as a short circuit and may be irreparably damaged. The non-polarised capacitor has no such polarity restrictions. Polarised capacitors are known as **electrolytic** capacitors and were developed to meet the demand for comparatively low voltage (up to about 450 V), inexpensive capacitors having a very large capacitance for their small size. This large capacitance to size ratio is achieved by using two aluminium plates separated by a conducting chemical solution (electrolyte). In manufacture, a d.c. voltage is applied to the plates and, as a result of electrochemical action, a thin film of aluminium oxide forms on one plate. Being an insulator, this aluminium oxide acts as the dielectric. The symbol for an electrolytic capacitor is shown in Fig. 7.6.

Fig. 7.6

7.6 Capacitive reactance

Capacitors offer a reactance to current flow and, as with inductors, this is frequency related.

$$\text{Capacitive reactance, } X_C = \frac{1}{2\pi fC} \text{ ohms,}$$

where f is the frequency in hertz and C is the capacitance in farads.

At d.c. or low a.c. frequencies, f is zero or very small, resulting in a large reactance, whereas at high a.c. frequencies, the reactance is small. Capacitors are often used as **d.c. blocking** devices. Capacitors are used for suppression or for preventing interference, **smoothing** d.c. signals and in timing circuits. These uses are discussed in later chapters.

In practice, the farad is a very large value of capacitance and it is more common for capacitors to be rated in microfarads (μF) or picofarads (pF).

8 The motor vehicle electrical system

8.1 Introduction

An electrical system consists of a number of components such as switches and motors connected together to perform one or more tasks. Although systems differ both in the components they contain and the jobs they are required to do, they operate on common electrical principles. A motor vehicle electrical system contains a wide range of components designed to perform a variety of tasks. Systems vary from vehicle to vehicle but each one carries out common tasks and, to a large extent, the system components differ only in size, shape and complexity.

8.2 Energy conversion and the vehicle electrical system

Energy is defined in Chapter 5 as the capacity for doing work, or alternatively, when energy is converted, work is done. To ensure that useful work is done, it is often necessary to convert energy from one form to another. In a motor vehicle, work is done when the road wheels are driven. The mechanical energy required to carry out this work is produced by the engine, which converts the chemical energy contained in petroleum to a mechanical form. Some of this mechanical energy available from the engine is in turn converted to electrical energy by a generator to energise the vehicle electrical system. Again, a portion of this electrical energy is converted into a chemical form and stored in the battery to be released as electricity during periods when the generator output is below a certain value.

The electrical system does work when energy is converted, for example to start the engine, light the lamps and operate the horn. All the components of the system which take energy from the generator or the battery are referred to as the **loads** of the system.

A vehicle electrical system can be simplified by considering it as consisting of a number of sub-systems, each one performing individual tasks.

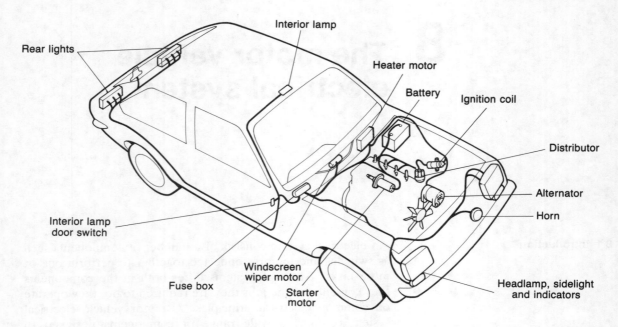

Fig. 8.1 Typical locations of electrical components in a car

The location of individual components of the electrical system varies from car to car. A typical layout is shown in Fig. 8.1.

8.3 Sub-systems

The electrical system can be regarded as consisting of six sub-systems.

a) The battery which stores energy for use by the electrical system;
b) The lighting sub-system (interior and exterior);
c) The starting sub-system which starts the engine until it runs under its own power;
d) The ignition sub-system which ignites the fuel in the engine;
e) The charging sub-system which provides power to the system and replaces the chemical energy transferred from the battery; and
f) The ancillaries sub-system which includes various operations such as the fuel gauge and horn.

Figure 8.2 illustrates the electrical system broken down into these sub-systems.

8.4 Circuits

A circuit is a complete closed path through which electricity can flow, and each sub-system can contain a number of circuits. Current flows from an electrical source to a load and

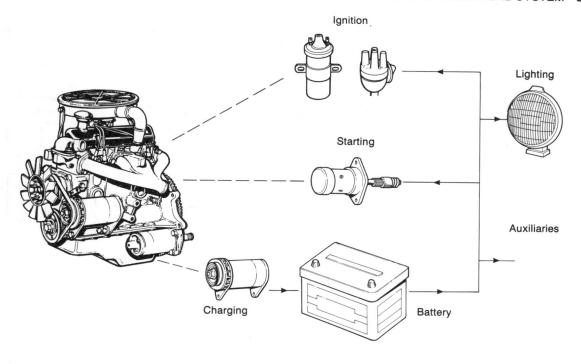

Fig. 8.2 The sub-systems

then back to the source and this is illustrated by the block diagram shown in Fig. 8.3.

The source of energy can be the battery or the generator and the load can be, for example, the horn. Conductors are usually copper wire covered with plastic or a rubber material. These **insulated** wires are used extensively but considerable savings in materials, and thus cost, can be made by using the vehicle metal chassis as the return conductor. In this case, the vehicle body is known as the earth return and the system is said to be an earth return system.

Protection of circuits is normally provided by **fuses**. Circuits can be controlled manually by switches or automatically by relays. Meters and gauges are used to monitor circuits and to

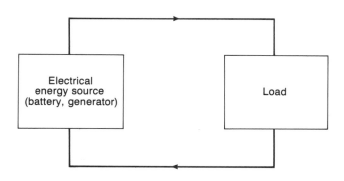

Fig. 8.3

provide the driver with a visual or audible check that the circuits are operating correctly. A block diagram incorporating this control, protection and indication is shown in Fig. 8.4.

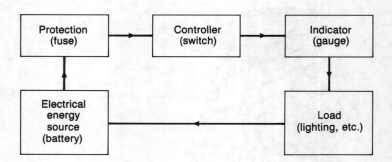

Fig. 8.4

9 The lighting sub-system

9.1 Introduction

Head, side, tail/brake and rear number plate lights are legal requirements for motor vehicles. When other lights are fitted, such as spot or fog lights, they must also satisfy the legal requirements regarding colour, position, power and the number fitted. All external lights together with the various interior lights constitute the lighting sub-system.

The electrical circuits comprise insulated wires which are normally bound together to form a **wiring harness**, or **loom**. This helps to protect against possible mechanical damage.

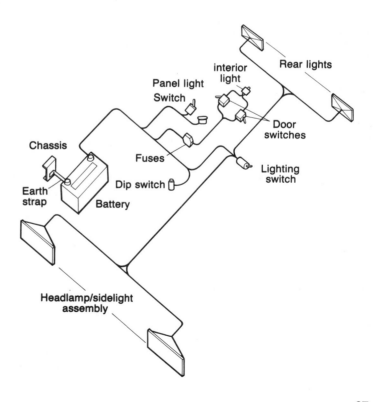

Fig. 9.1

Fuses are usually used to protect the circuits from the effects of electrical overloads or **short circuits**.

Lamps basically consist of a bulb (the light source), which is mounted in front of a reflector. These are covered by a plastic or glass lens for protection and for shaping the light beam.

The power rating, or **wattage**, of bulbs varies in size dependent upon the bulb's usage. A headlamp will have a higher wattage than an interior panel light.

Control of the lights is achieved by various switches fitted to the dashboard or the steering column, or, indeed, in the case of interior lights, by door switches which operate when the door is open. Brake lights operate by means of a brake switch controlled by the operation of the foot brake.

9.2 Lighting circuits

Lighting circuits are generally connected in parallel across the battery terminals. Where more than one bulb is used in a circuit then these are in turn connected in parallel with each other. The number of fuses varies from vehicle to vehicle. For ease of understanding the circuitry, Fig. 9.2 shows only the interior light circuit with a fuse connected but all the other circuits without fuses.

Series connected switches control the circuits and, when closed, allow current to pass through the load (bulb). The heating effect of the current then causes the bulb filament to emit light when the temperature of the filament is large enough.

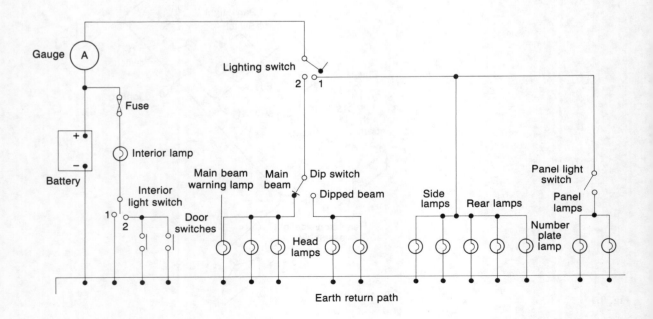

Fig. 9.2 Circuits of the lighting sub-system

The headlamps are not only controlled by the lighting switch but can also be switched from dipped to main beam and back again by a dip switch.

9.3 Light bulbs

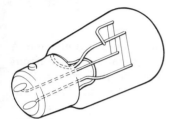

Fig. 9.3

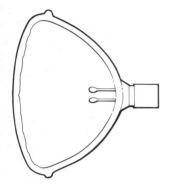

Fig. 9.4

These generally consist of a sealed glass envelope containing a tungsten filament which is surrounded by an inert gas or by a vacuum. Some bulbs contain two separate filaments, e.g. the headlamp bulb. One end of each filament is connected to the metal part of the bulb cap and the other end to its own contact in the insulated end of the bulb cap.

Most headlamps are **sealed beam units** containing the two filaments for main and dipped beam. These are not bulbs as such and the unit has a shaped front which acts as the lens and a silvered reflective back and also protects against damp or dirt.

Bulbs are classified according to their voltage, power rating and end cap construction. A typical side light bulb could well be classified as a 12 V, 6 W bayonet type. The power rating gives an indication of the amount of light that the bulb produces. Examples of the common types of end cap construction are shown in Fig. 9.5.

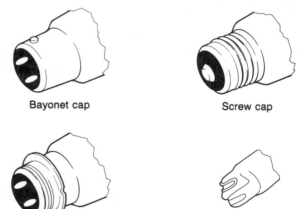

Bayonet cap   Screw cap

Prefocussing flange   Wedge base

Fig. 9.5

9.4 Circuit protection

Heat energy is produced when an electric current flows and this raises the temperature of the conductor. If components are added indiscriminately to a circuit or a short-circuit fault occurs, then the circuit resistance falls and the current will be excessive. The resulting rise in temperature can damage components and burn insulation.

A short-circuit fault is defined as the instant when two points of opposite polarity come into direct contact with each other. The damaging effects of this can be extreme. For example, assume that a short-circuit fault occurs directly between the terminals of a 12 V battery and that the battery internal resistance is 0.05 ohms. The resulting current flow will be

$$I = \frac{12(\text{V})}{0.05(\Omega)} = 240 \ (\text{A}).$$

To protect circuits from these effects, fuses are connected in series with the load. They are made of metal and are current rated. If a current flows which exceeds the fuse rating, then the fuse melts or **blows** and creates an **open circuit**. The circuit is opened and no current can flow.

Fuses must be selected such that their rating is slightly higher than the maximum circuit current expected to flow. A circuit taking 12 A can be protected by a 15 A fuse.

Whenever a fuse fails then the cause of the failure should be investigated. It may have failed due to age or due to a temporary accidental overload, but on the other hand it can be due to a possible serious fault. Fuses should always be replaced by another fuse of the same rating.

9.5 Conductor colour coding

Table 9.1 Conductor colour code

Colour	Number
Black	0
Brown	1
Red	2
Orange	3
Yellow	4
Green	5
Blue	6
Violet	7
Grey	8
White	9

The various conductors used in the system are normally grouped together into a loom or harness, for protection. The individual conductors themselves are colour coded to provide ease of recognition. This colour coding is numerically similar to the resistor colour code and is shown in Table 9.1.

Using the colour code, coloured cables can be represented on black and white circuit diagrams. For example, a red cable with a white tracer would appear on a wiring diagram as shown in Fig. 9.6. It must be noted that this colour code is not applied universally.

When conductors are connected directly to terminals, then cable connectors are used. Examples of commonly used connectors are illustrated in Fig. 9.7.

Fig. 9.6

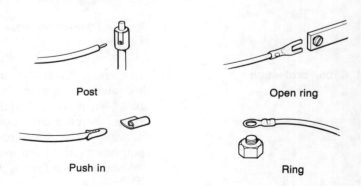

Fig. 9.7

Post

Open ring

Push in

Ring

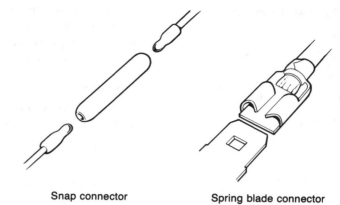

Fig. 9.8

Snap connector   Spring blade connector

Cable-to-cable connections are often made with snap connectors or spring blade connectors (Fig. 9.8).

9.6 Switches and switch classification

(a) Single pole, single throw switch

(b) Single pole, double throw switch

(c) Double pole, four position (or four way) switch

Fig. 9.9

Circuits can be opened or closed using manually operated switches which have one or more movable contacts. Switches normally include a spring mechanism to give a fast action. In the closed position, this spring ensures that the contacts make good electrical contact and when opened, the fast action reduces any damage due to **arcing** between the contacts. Switches are rated according to the maximum current they control and the maximum voltage which is applied to them. The material used for the contacts depends on the intended function of the switch. For general purposes, phosphor bronze is used, whereas silver- or gold-plated contacts are often found in more sophisticated equipment such as computers. As a rule, switches are connected into circuits on the **live** or high-voltage side of the load. This ensures that components are isolated from the supply when the switch is open and can therefore be handled safely.

A common way of describing various switches is by referring to the number of positions in which the switch can be closed and to the number of movable contacts or poles of the switch. Three examples are shown in Fig. 9.9. Additional information can also be gained from certain switch symbols (Fig. 9.10).

Figure 9.10*a* illustrates a switch that has a central **off** position compared to Fig. 9.10*b* which is a **changeover** switch. The bar on the movable contact of the switch in Fig. 9.10*b* indicates that this is a make-before-break switch and that a second circuit is made before the first circuit is broken. Figure 9.9*b* shows a break-before-make switch.

9.7 Equivalent circuits

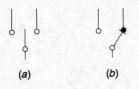

(a)   (b)

Fig. 9.10

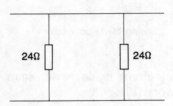

24Ω   24Ω

Fig. 9.11

In order to simplify calculations and to make circuit operations easier to understand, each of the circuits in the lighting sub-system can be replaced by an equivalent circuit. The bulb filaments are replaced in the equivalent circuit by resistors.

Example 9.1

A circuit has two side lights and each bulb is rated at 12 V, 6 W. The working resistance of each filament will be

$$P = \frac{V^2}{R} \text{ watts.}$$

Therefore,

$$R = \frac{V^2}{P} \text{ ohms}$$

$$R = \frac{12 \times 12}{6} = 24 \text{ ohms.}$$

The equivalent circuit to the two bulbs in parallel is now two 24 Ω resistors in parallel (Fig. 9.11).

These resistors can in turn be represented by a single resistor of 12 Ω, since

$$R_T = \frac{R_1 \times R_2}{R_1 + R_2}$$

as shown in Chapter 4 i.e.

$$R = \frac{24 \times 24}{24 + 24} = 12 \ \Omega.$$

The whole of the sub-system can be represented by an equivalent circuit consisting of a single resistor connected to the battery terminals by extending the process of Example 9.1.

10 The starting circuit

10.1 Insulators and conductors

The conducting elements of the starting sub-system consist of insulated cables, braided earthing straps and the vehicle chassis.

1   The insulated cables comprise copper wire strands wound into lengths and covered by rubber insulation or plastic material. The insulation provides protection against possible short-circuit damage.
2   Flexible copper earthing straps are used and these are braided to form flat, wide conductors.
3   The vehicle chassis provides an earth return from circuit components to the battery. The starter motor and solenoid switch are usually earthed through engine mounting bolts to the engine, which in turn is earthed to the chassis by an earthing strap.

The resistance of conductors is determined by their physical dimensions and their **resistivity** (inherent resistance). Metals have a low resistivity and are generally good electrical conductors. Insulating materials, such as rubber and polyvinyl chloride (PVC), have a high resistivity and do not conduct electricity. Some types of varnish are also good insulators and are often used to insulate wires used for making coils.

The most common conductor of electricity, and the one that is most widely used in electrical circuits, is copper. Aluminium and steel are also used as conductors although they have a higher resistivity than copper. When choosing conducting materials, circuit designers must consider factors such as size, weight, strength, cost and temperature.

The length of cable required and the circuit current determine the diameter of the conducting cable. This is due to the principles of Ohm's law and the fact that the resistance of a conductor is proportional to its length and inversely proportional to its cross-sectional area. Consequently, cables required to carry large currents have larger diameters than those carrying smaller currents.

10.2 The starting sub-system

The components of the sub-system are illustrated in Fig. 10.1. Operation of the ignition starter switch energises the solenoid switch and completes the starter motor circuit. The armature of the starter motor rotates, and a pinion gear mounted on the armature is forced into engagement with gear teeth fitted around the engine fly wheel. Due to the electrical energy input, the engine rotates to a speed at which the energy supplied by the fuel is sufficient to keep it rotating without any electrical energy input, that is, the engine starts. The pinion automatically withdraws from the gear teeth and this prevents any back drive to the starter motor from the engine.

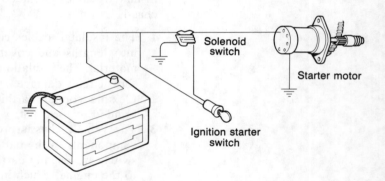

Fig. 10.1

Large currents, in the order of 250 to 350 amperes, are drawn from the battery on starting to produce the large amount of energy required to rotate an engine.

The sub-system is divided into two main circuits.

The control circuit

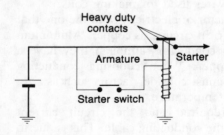

Fig. 10.2 The control circuit

This is illustrated in Fig. 10.2. When the starter switch is operated current flows through the coil and the soft iron armature is attracted to the centre of the coil due to electromagnetic effects. Once the heavy-duty contacts are made, the starter motor circuit is completed. The armature will return to its original position by spring action when the starter switch is released.

The starter motor circuit

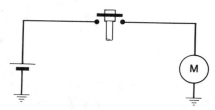

Fig. 10.3 The starter motor circuit

When the circuit (Fig. 10.3) is complete, current flows from the battery, through the solenoid switch contacts and through the motor to earth. The motor converts the electrical energy into mechanical energy by utilising the magnetic effect produced when an electric current flows through a coil. A turning force, or **torque**, is produced and this turns the motor armature.

10.3 The solenoid switch

A solenoid switch consists of a coil of approximately 400 turns of insulated copper wire wound on an iron core armature. Figure 10.4 shows the exploded view of a typical solenoid switch.

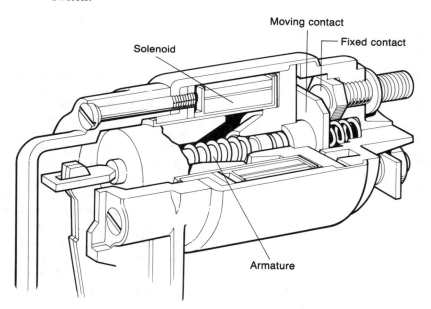

Fig. 10.4

It is usual for the switch to be operated by a smaller starter switch which is part of the ignition switch. The large currents carried by the solenoid switch create large voltage drops in the cable and these can be reduced if the battery, solenoid switch and starter motor are fitted close together to reduce the length of cables required.

10.4 The starter motor

The starter motor (Fig. 10.5) converts electrical energy to mechanical energy. It is a **series** wound machine in that the field windings are connected in series with the motor

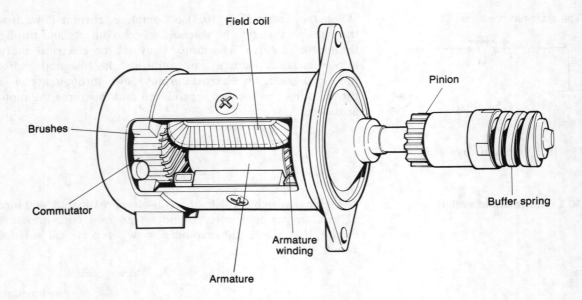

Field coil

Pinion

Brushes

Commutator

Armature winding

Armature

Buffer spring

Fig. 10.5

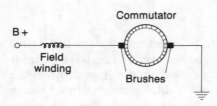

Fig. 10.6

armature. Its characteristics are such that it develops a maximum torque initially to turn the engine over, the torque decreasing quickly as the engine runs up to speed. It is essential that this type of motor is operated on load, otherwise its speed will continue to increase and damage will occur.

The motor body, or **yoke**, is a steel casing which carries **four pole shoes**. Copper or aluminium field coils are wound around each shoe. The **armature**, which is free to rotate between these shoes, consists of a steel shaft fitted with soft iron stampings, or **laminations**. Slots are cut into the laminations and these cuts house insulated copper strips to form the armature winding coils. These coils, together with the field coils, act as two electromagnets and the magnetic principles of attraction and repulsion between the two magnets cause rotation of the shaft.

A **commutator** (Fig. 10.6), consisting of copper segments which are insulated from each other and from the motor shaft, is provided with two carbon **brushes** which are forced into contact with the commutator by means of springs. When the starter switch is closed current flows from the battery through the field windings and through the armature coils that are in contact with the brushes. The electromagnetic fields which are produced cause sufficient force to rotate the armature. The commutator begins to move, and as pairs of segments come into contact with the brushes, the armature coils are connected in turn to the supply, maintaining continuous rotation.

Maximum current flows when the motor is first energised. As speed increases an e.m.f. is induced in the armature windings, the polarity of which opposes the supply voltage, thereby reducing circuit current. When running at full speed, the current drawn from the supply is just sufficient to overcome friction and other inherent motor losses.

11 Vehicle ignition system

11.1 The petrol engine

The power to move a vehicle is provided by the engine which produces the required mechanical energy from the chemical energy of the fuel. The mixture of petrol and air in the engine **cylinders** must first be ignited, and this is achieved by the electrical ignition circuit. In a four-cylinder engine, as shown in Fig. 11.1, four **pistons** are connected to the **crankshaft**. When the starter motor is operated, the crankshaft rotates and the pistons move up and down in the cylinders. The crankshaft also drives the distributor and fuel inlet and exhaust valves. The distributor controls the high-voltage ignition pulses by

Fig. 11.1 Four-cylinder engine

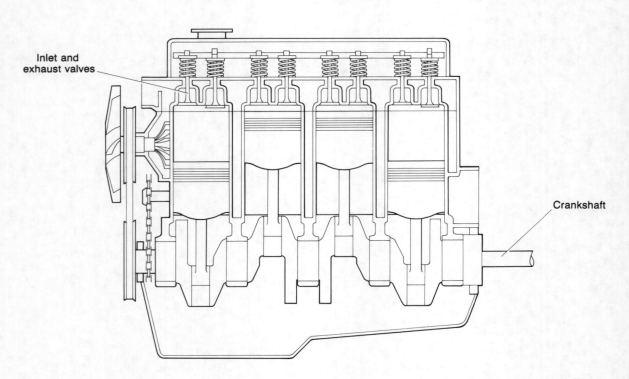

Inlet and exhaust valves

Crankshaft

feeding them to each cylinder in turn. Once the fuel is lit, the expanding gases force the pistons up and down maintaining the crankshaft rotation and the engine runs under its own power.

11.2 The ignition circuit

A block diagram of the ignition sub-system is shown in Fig. 11.2. The system can essentially be broken down into two circuits:

1   the **low tension (LT) primary circuit**, which includes the **primary winding** of the ignition coil and the **contact breaker**; and

2   the **high tension (HT) secondary circuit**, which includes the **secondary winding** of the **ignition coil**, **rotor arm**, **sparking plugs** and **HT connecting leads**.

When the contact breaker points are closed and the ignition switch is operated, current flows through the primary circuit. This current sets up a magnetic field around the primary and secondary windings. The distributor shaft is mechanically linked to the crankshaft and as the engine is turned over, the shaft turns and the **cam** forces the spring-loaded points apart.

Fig. 11.2

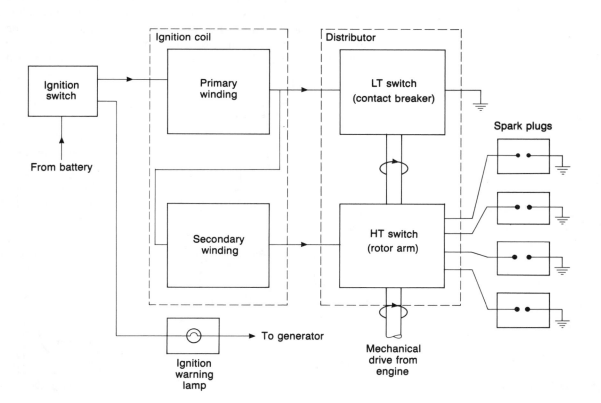

Current in the primary circuit ceases to flow, causing the magnetic field to collapse. This changing magnetic field induces a very high voltage into the secondary winding. The process repeats itself continuously as the points make and break, thereby inducing a series of high-voltage pulses into the secondary winding.

The secondary winding is connected to the rotor arm through an HT lead and a carbon brush. The rotor arm rotates on the distributor shaft, and as it moves past a fixed electrode, the voltage pulse induced in the secondary overcomes the air gap between the arm and the electrode and also the air gap between the sparking plug electrodes, causing current to flow through the secondary circuit. This current produces a spark across each plug's electrodes to earth as the rotor arm rotates.

The speed at which the magnetic field collapses and the turns ratio between secondary and primary windings determine the value of the induced voltage. This HT voltage can be in the order of 20 000 V.

A capacitor (or condenser) is connected between the points and prevents excessive arcing or sparking across them as they open. It also assists in causing a more rapid collapse of the magnetic field.

When the starter motor is energised, the current taken from the battery is at a maximum and the battery voltage falls below its normal value. This reduction in battery voltage reduces the operating voltage to the coil and consequently reduces the induced voltage in the secondary winding. The performance of

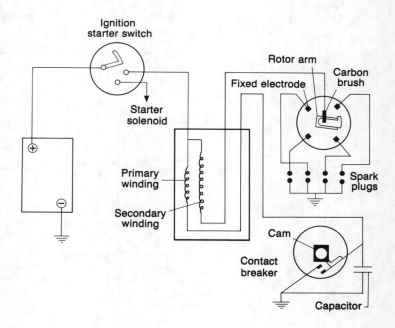

Fig. 11.3 Ignition circuit – schematic

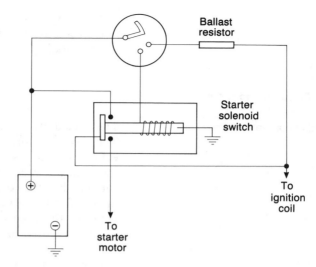

Fig. 11.4

the ignition circuit is reduced and, particularly in cold weather, may prove unsatisfactory. A **ballast resistor** connected in series with the ignition switch and the primary winding, as shown in Fig. 11.4, allows the use of a coil that has a lower operating voltage than the unloaded battery terminal voltage.

Operation of the starter switch closes the solenoid switch to energise the starter motor. The ballast resistor is short circuited and the reduced battery voltage is applied to the coil. When the starter switch is released, the ballast resistor is connected in circuit with the coil and forms a potential divider circuit with the primary winding. The value of the ballast resistor is usually designed to be equal to the resistance of the primary winding.

With a 12 V battery system and where R_b is the ballast resistance, and R_p is the primary coil resistance, then, for normal running, the coil voltage is

$$V_c = 12 \times \frac{R_p}{R_b + R_p}$$

$$= \frac{12}{2} \ (R_b = R_p)$$

$$V_c = 6 \text{ V}.$$

A coil designed to operate at 6 V is used.

Assume that during starting conditions the battery voltage falls to 10 V. Because R_b is short circuit, then

$$V_c = 10 \text{ V}.$$

Although this voltage is above the coil operating value, it only occurs for a few seconds and the coil is not damaged due to overheating.

11.3 Circuit components

The distributor and timing

The breakdown of a typical distributor is shown in Fig. 11.5. The cover is normally made of plastic and houses the fixed electrodes and a spring-loaded carbon brush. The rotor arm is essentially a metal contact fitted to a plastic block which is rotated by the distributor shaft. The contact breaker is mounted on a plate which allows for adjustment of the gap between the points as well as of the position of the contact breaker relative to the cam for timing purposes.

It is essential for correct engine running that fuel ignition in the cylinders occurs in sequence. This is known as the firing order. The time at which ignition takes place in relation to the position of the pistons is also important. This timing varies with the engine speed and the load placed upon it. As speed increases, ignition generally occurs earlier in the cycle of engine operations or **advances**. The advance mechanism in the distributor automatically performs this function to ensure efficient operation of the engine.

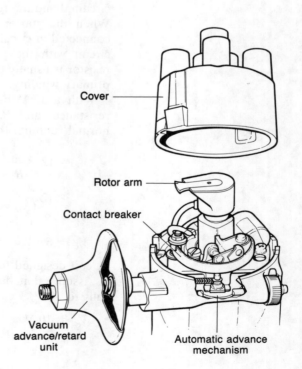

Cover

Rotor arm

Contact breaker

Vacuum advance/retard unit

Automatic advance mechanism

Fig. 11.5 Distributor

Sparking plugs

These provide the spark to ignite the fuel and air mixture in the engine cylinders. A plug (Fig. 11.6) consists of two electrodes which are insulated from each other, one being connected to earth through the engine, and the other to the secondary winding of the ignition coil. The high resistance of the air gap between the electrodes breaks down when a large potential difference exists across the gap. This process, known as **ionisation**, causes the air gap to behave like a conductor. A spark jumps between the electrodes.

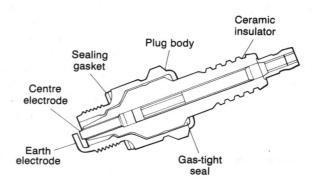

Fig. 11.6 Sparking plug

Ignition coil (Fig. 11.7)

The primary and secondary windings are both wound on a laminated iron core and sealed in a metal container. The secondary windings consist of many turns (typically 20 000) and these are wound in layers, each layer being insulated from the others by paper. The primary windings are wound with heavier gauge wire than the secondary winding and are usually

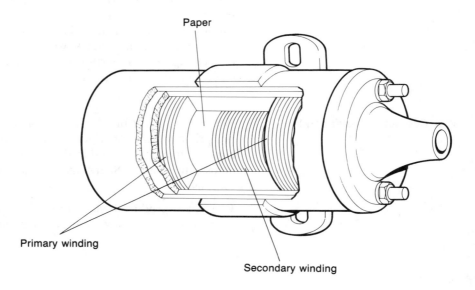

Fig. 11.7 Ignition coil

in the order of 300 to 400 turns. Connections to the primary winding for the LT circuit are by two terminals at the top of the coil. One end of the secondary is connected to the primary winding and the other to the iron core, which is in turn connected to the centre of the coil top (HT terminal).

Energy losses, due mainly to heat, are reduced by the laminated construction of the core and by filling the container with an insulating oil.

11.4 Tracking

The expression **tracking** refers to low-resistance leakage paths across insulators resulting from an accumulation of moisture and/or dirt. Tracking can occur in the ignition system between the engine and the insulated terminal of the sparking plug or between the rotor arm and the distributor shaft. Current will track along these leakage paths due to the high voltage of the secondary circuit and this will result in partial, or complete, failure of the ignition system.

11.5 Interference and suppression

Electromagnetic induction in the ignition system can cause unwanted voltages to be induced in other electrical circuits, leading to interference. This is particularly the case with radio or television receivers in the vicinity.

Extra resistance is added to the secondary circuit in the form of high-resistance HT cables or by connecting special resistors (**suppressors**) in the HT leads. This effectively reduces the current and therefore the strength of the magnetic field through the circuit when the air gap breaks down, so reducing any interference.

Capacitors can also be connected between areas of possible interference and earth to improve suppression. Further suppression can be achieved by enclosing components in metal cases and covering cables with braided metal sheaths and connecting them to earth. This is called **screening** and prevents electromagnetic energy radiating from the circuits.

11.6 Contact breaker sparking

When the contact breaker points open, an e.m.f. is induced in the primary winding which attempts to maintain the primary circuit current. This induced current results in a large spark across the points. Sparking is undesirable for the following reasons.

1 Heat is produced which effectively burns and damages the points, in turn developing a high surface resistance.

2 Metal transference takes place across the points causing one contact to become **pitted**, and the other to be spoiled with deposits of metal.

3 Energy is wasted because sparking delays the collapse of the magnetic field around the primary winding, thus reducing the e.m.f. induced into the secondary winding.

By connecting a capacitor across the points, sparking is reduced. Induced current flows into the capacitor when the points open instead of creating the sparking across them. This capacitor is often termed a **condenser**.

12 d.c. generators

12.1 e.m.f. generation and flux cutting

The principles of electromagnetic induction due to flux linking are discussed in Chapter 6. Another way in which electromagnetic induction occurs is by a process known as **flux cutting**. When a bar magnet is moved inside a solenoid which is in turn connected to an electrical circuit, then an e.m.f. is induced into the circuit. It is immaterial which moves, the magnet or the conductor. A d.c. generator uses the principle of flux cutting for its operation by rotating a conductor between the poles of a magnet. The magnitude of the induced e.m.f. depends on:

1 the strength of the **magnetic flux**;
2 the number of turns of wire in the coil; and
3 the rate of relative motion between the conductors and the flux.

It is important to know the polarity of the induced e.m.f. and the direction in which current will be made to flow through a circuit. This can be found using Fleming's right-hand rule. Figure 12.1 illustrates this rule, where:

1 the thuMb indicates the direction of Motion of the conductor;
2 the First finger indicates the direction of the Flux; and
3 the Second finger gives the Sense or direction of the induced e.m.f.

Fleming's right-hand rule

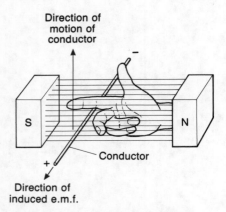

Fig. 12.1 Fleming's right-hand rule

12.2 Magnetic flux

The unit of magnetic flux Φ is the **weber** (Wb) and this can be stated as that magnetic flux which generates an e.m.f. of one volt when cut by a conductor in one second. The **flux density** (B) of a magnetic field is measured in webers per square metre, which is represented by the **tesla** (T).

When flux is cut at a constant rate, the e.m.f. in volts is equal to

$$\text{e.m.f.} = \frac{\text{webers}}{\text{seconds}}.$$

When a conductor of length l moves at right-angles to a uniform magnetic field of flux density B teslas through a distance of d metres then the total flux cut will be Bld webers.

In t seconds, the induced e.m.f. is:

$$\text{e.m.f.} = \frac{Bld}{t} \text{ volts}.$$

12.3 The elementary generator

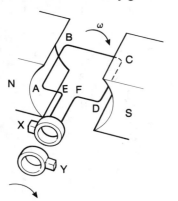

Fig. 12.2

When a rectangular loop is rotated, in the direction shown in Fig. 12.2, between the poles of a magnet then:

1 no e.m.f. will be induced in sides BC or AE and DF, since these are normally situated outside of the magnetic field and no flux is being cut; and

2 the e.m.f. induced in AB is opposite in polarity to that induced in CD as they are each cutting the flux in opposite directions, i.e. when AB is moving upwards, CD is moving downwards. These induced voltages are therefore effectively additive and the total e.m.f. is equal to the sum of the two induced voltages.

By connecting the loop to two **slip rings** which rotate with the loop, the electrical potentials produced by rotation can be sensed by **carbon brushes** (X and Y) which are forced into contact with the slip rings. This output voltage will follow a **sinusoidal** path and is a form of alternating current (a.c.), as shown in Fig. 12.3.

When the loop is in the position shown in Fig. 12.2, AB is moving upwards and CD downwards at right-angles to the lines of flux and the rate at which it is cut is a maximum. The induced e.m.f. therefore is a maximum value. After the loop has rotated 90° then the direction of rotation of the turn is parallel to the flux. No flux is cut and the induced e.m.f. is zero. A further rotation of 90° will cause a maximum e.m.f. to be induced of the opposite polarity because AB is now moving downwards and CD upwards. This is illustrated in Fig. 12.3.

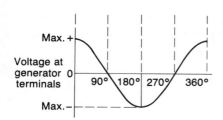

Fig. 12.3

12.4 Commutator action

The a.c. sinusoidal output of the generator can be converted to d.c. by using a **commutator** instead of slip rings. In its simplest form, a commutator (Fig. 12.4) consists of a single ring split into two segments which are insulated from each other. Its action is to reverse the connection to the external load R such that the left-hand brush is always in contact with the side of the loop which is positive and the right-hand brush with the side of the loop which is negative. Current flow through R is always in the same direction.

Smoother outputs are produced in practice using armatures that consist of many insulated coils wound on a laminated steel core which is built up on a drive shaft. Multi-segment commutators are used and each coil shares one segment with another coil. All the coils are therefore linked together in series-parallel paths to form a complete **armature winding**.

The magnetic field in which the armature is made to rotate is sometimes produced using permanent magnets. However, it is more common to produce the field electromagnetically by **field coils** which are energised by a direct current. These field coils are usually pre-formed to fit around iron or steel pole shoes or **pole pieces** which are fixed to the frame of the

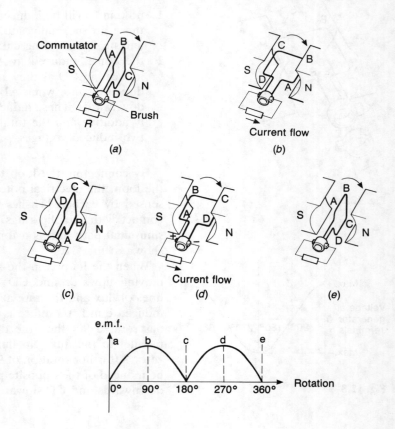

Fig. 12.4 Commutator action

machine. The shoes are evenly spaced and the coils are connected together to form a **field winding** in such a way that alternate north and south poles are produced.

At least two pole shoes must be used in any machine but a stronger and more uniform magnetic field is produced using more than one pair. It is common practice to fit machines with as many brushes as there are poles.

12.5 Generator types

Field excitation generators

The magnitude of the induced e.m.f. of a given machine is directly proportional to the armature speed and the strength of the magnetic field. When field coils are used to produce the magnetic field then over a limited range of excitation currents, its strength, and therefore the induced e.m.f., is proportional to the excitation current passing through the field winding. The field windings can be connected to the armature so that an e.m.f. induced in the armature passes current through the field.

Alternatively, field excitation can be achieved by an external current.

Self-excitation generators

Self-excited generators rely on pole shoes possessing some **residual magnetism**. On starting, a small induced e.m.f. due to residual magnetism in the pole pieces passes current through the field winding. This causes the magnetic field to increase in strength so increasing the magnitude of the induced e.m.f. More current flows in the field and this process continues until the induced e.m.f. reaches the rated value for the speed at which the machine is running. The field windings for self-excited generators are said to be **series wound**, **shunt wound** or **compound wound** according to the way in which the field and armature are connected.

Series-wound generator

This type of generator has its field winding connected in series with the armature, as shown in Fig. 12.5. The machine must be placed on-load to allow voltage buildup due to residual magnetism. Because the field carries the full load current, the field coils are of heavy-gauge wire. The large current that passes through the coils means that the ampere turns required to produce a strong magnetic field are obtained by using relatively few turns to form the coils.

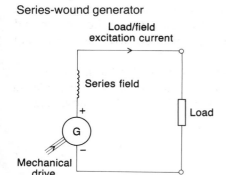

Fig. 12.5 Commutator action

Shunt-wound generator (Fig. 12.6)

This has its field connected in parallel with the armature. Residual magnetism induces an e.m.f. in the armature and the generator drives up to its operating speed and the terminal voltage builds up to its rated value. The field is designed to have sufficient resistance to limit the power absorbed in the field to three or four per cent of the generator output. The field coils, therefore, comprise many turns of fine wire. The large number of turns ensures that a strong magnetic field is produced even though the excitation current is relatively small.

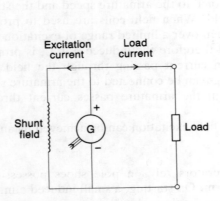

Fig. 12.6

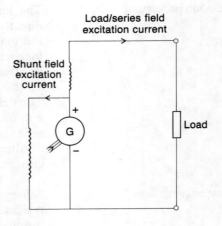

Fig. 12.7

Compound-wound generator (Fig. 12.7)

The magnetic field in a compound-wound generator is due to both a series and a shunt winding. When the generator is off-load, the e.m.f. builds up due to the shunt field. When the machine is on-load, the load current excites the series field. The series field normally aids the shunt field and this strengthens the magnetic field as the load is increased.

Separately excited generators (Fig. 12.8)

These do not rely on residual magnetism for their operation. The field winding is energised by an external d.c. source and there is no electrical connection between the field and armature circuits.

The generated voltage achieves its final value much quicker than with self-excited machines but the need to provide an external d.c. supply for field energisation is a disadvantage.

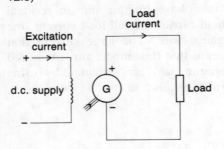

Fig. 12.8

12.6 Generated e.m.f. and terminal voltage

The armature circuit of a d.c. generator can be considered to consist of a voltage generator connected in series with a resistor. In Fig. 12.9 E_g represents the output of the voltage generator (the e.m.f. induced in the armature winding), R_a represents the combined resistance of the armature winding, commutator and brushes, I_a represents the armature current, and V represents the voltage at the armature terminals.

Applying Kirchhoff's voltage law to the circuit

$$E_g = I_a R_a + V$$

or, by rearranging this equation,

$$V = E_g - I_a R_a.$$

Because the armature circuit resistance is usually small, $I_a R_a$ is small. Therefore, the difference between the applied voltage and the generated e.m.f. is small.

Example 12.1

What is the generated e.m.f. for a d.c. generator required to produce a terminal voltage of 225 V when delivering 90 A? The armature resistance of the machine is 0.1 Ω.

The p.d. due to armature resistance is:

$$I_a R_a = 90 \times 0.1 = 9 \text{ V}$$

$$E_g = I_a R_a + V$$

$$= 9 + 225$$

$$E_g = 234 \text{ V}.$$

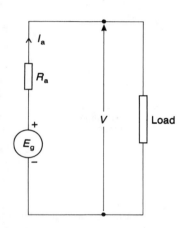

Fig. 12.9

12.7 Generator power conversion

Off-load conditions

When a generator is operating off-load, then the mechanical power source provides power equal to the losses incurred in driving the armature at the speed required to achieve the rated voltage of the generator. In a car, the mechanical power source is the engine.

On-load conditions

Under loaded conditions, the armature current produces a torque which opposes the motion of the armature. Because the armature is therefore harder to turn, the torque from the engine must increase and this will lead to a reduction in engine speed. A **governor** is generally used to control the engine speed. Any changes in engine speed are sensed by the governor, which alters the fuel intake to keep the engine running at the required speed.

Any changes in electrical load that produce changes in the

torque opposing armature current produce corresponding changes in the mechanical power drawn from the engine.

12.8 Voltage control

The generator output voltage is usually controlled by varying the field excitation current. An example using a field-regulating variable resistor for a separately excited generator is shown in Fig. 12.10.

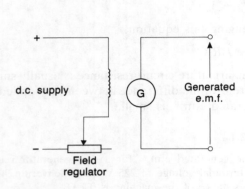

Fig. 12.10   Fig. 12.11

The open-circuit, or off-load characteristic, for a generator running at constant speed showing the relationship between field current and generated e.m.f. is illustrated in Fig. 12.11.

A small e.m.f., V_{res}, may be produced for zero field current due to residual magnetism. A point B is eventually reached where further increases in field current produce no changes in the generated e.m.f. At this point the pole pieces of the generator are said to be **magnetically saturated** when the amount of magnetic flux developed in the material has reached its limit. The generator voltage is therefore controlled between points A and B, as shown in Fig. 12.11 when small changes in field current produce large changes in the generated e.m.f.

12.9 Load characteristics

These illustrate the manner in which the generator terminal voltage changes for changes in load. The generator is driven at a constant speed and the terminal voltage is measured for decreasing load resistance values which will, in turn, increase the armature current.

Separately excited generator (Fig. 12.12)

The terminal voltage decreases as the load current increases. This is mainly due to the I_aR_a voltage drop which occurs as the load current increases. The field-regulation resistor discussed in Section 12.8 provides a sensitive control of the generator output and makes this type of generator suitable for use in motor control systems.

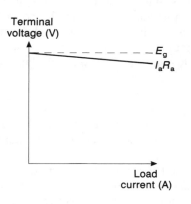

Fig. 12.12

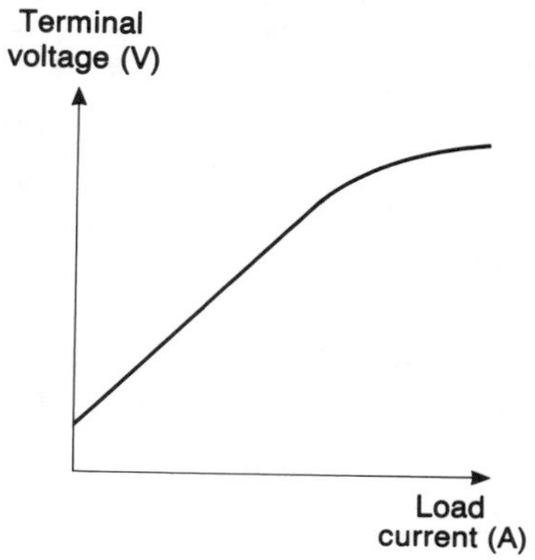

Fig. 12.13

Series generator (Fig. 12.13)

Increases in load current cause increases in the magnetic field produced in the field winding because the load current passes through the field winding. The generated e.m.f. increases and, consequently, the terminal voltage increases to a point when magnetic saturation is reached. The field current can be controlled partially by connecting a **diverter** resistor in parallel with the field, but this does not maintain a constant terminal voltage. For this reason, the series generator is only used in special applications.

Shunt generator (Fig. 12.14)

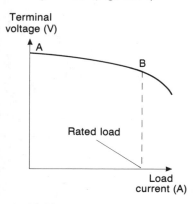

Fig. 12.14

Initially, terminal voltage falls due to the I_aR_a voltage drop. However, the field current I_a depends on the terminal voltage and, consequently, on the magnetic field, and therefore the terminal voltage tends to decrease for any increase in load current. Between points A and B this reduction in terminal voltage is fairly small, but if the load current is increased beyond point B then it decreases rapidly due to the **shunting** effect of the load on the field circuit. By using a shunt generator for maximum loads corresponding to point B, the terminal voltage can be controlled by a field-regulating resistor connected in series with the field, providing that only slow changes in load current occur. This is ideal for application such as battery charging where a constant voltage is required over a wide range of load currents, and where the load does not change rapidly.

Compound generator (Fig. 12.15)

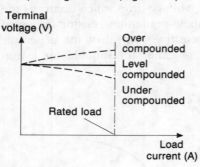

Fig. 12.15

In most compound generators the series field is wound so that it aids the shunt field and the machines are said to be **cumulatively compound**. They are designed so that the rise in terminal voltage due to the series winding compensates for the I_aR_a voltage drop due to the shunt winding. This is achieved by the turns ratio between the series and shunt fields and produces machines that are said to be **over, level** or **under compounded**. The output voltage can also be controlled using a field regulator connected in series with the shunt field. Level compounded machines are ideal where constant voltage supplies are required, regardless of rapid and large changes in load.

12.10 Output voltage polarity change

To change the polarity of the terminal voltage the direction of rotation of the armature, or the direction of the magnetic field, must be reversed.

This is easily achieved for separately excited generators but is difficult in self-excited machines as they rely on residual magnetism for the buildup of the generated e.m.f. The magnetic field due to residual magnetism has a set direction. Attempts to reverse the output polarity tend to weaken the residual field, thereby preventing the buildup required.

13 The charging sub-system

13.1 Introduction

A generator provides the power required by a vehicle electrical system by converting some of the mechanical energy produced by the engine into electrical energy. The electrical energy supplies the system directly and maintains the battery in a charged condition. When the engine is stationary, the battery provides the power supply.

The electrical voltage, and hence the current output from the generator, rises as the engine speed increases and some form of current–voltage control is required to avoid damage to components and to ensure the most efficient use of the available energy.

The components that produce a controlled supply of electrical power form the charging system and this is illustrated in block form in Fig. 13.1.

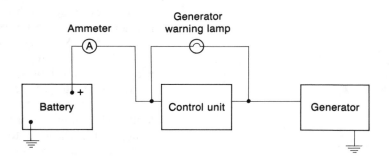

Fig. 13.1 The charging sub-system

13.2 Circuit operation

A schematic diagram of the charging sub-system is shown in Fig. 13.2. The control unit consists of three relays, **the cut-out**, the **current regulator** and the **voltage regulator**. The generator output terminals are marked D+ and D− and the field winding terminal as F.

The **ignition/generator warning lamp** will light when the ignition switch is operated, by current passing from the

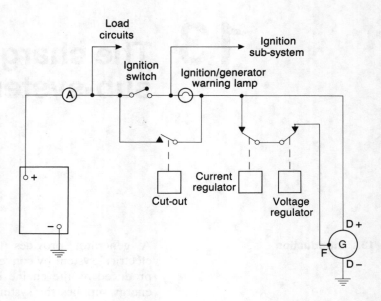

Fig. 13.2

battery, through the ignition switch and warning lamp and to earth through the generator.

A **centre zero** ammeter indicates the rate of battery discharge due to the ignition sub-system and any other loads which are in operation at this time.

When the engine starts, the generator produces a direct output voltage which opposes the battery voltage thereby reducing the current flow through the warning lamp. As the engine speed increases, the generator output voltage increases and the lamp is extinguished. When the generator voltage exceeds the battery voltage, the cut-out operates and connects the generator to the battery and other load circuits. For what is called a 12 V system, the cut-out is designed to operate at about 13 V. The generator warning lamp is bypassed and is prevented from lighting while the cut-out contacts are closed. As the generator output continues to rise, current now flows from the generator to charge the battery and supply other loads. The ammeter now reads in the opposite direction and indicates the charge rate of the battery.

The current and voltage regulator contacts open and close continuously and interrupt the field circuit. If the generator current rises above a predetermined value, the current-regulator contact opens, reducing the field current and therefore the generator output current. The voltage-regulator contact operates in a similar manner. Typical values of the maximum current and voltage are 30 A and 15 V, respectively.

Whenever the generator voltage falls below the battery voltage, the cut-out contacts open and increasing current from the battery lights the generator warning lamp. This provides a visual warning to the driver which indicates a fault in the charging system. The speed of the engine is normally set so that the generator warning lamp remains extinguished at **tick-over**. Many vehicles are not fitted with an ammeter and the warning lamp is the only indication of fault conditions.

13.3 The generator

The charging sub-system uses a self-excited shunt-wound generator, as explained in Chapter 12. An exploded view illustrating typical generator construction is shown in Fig. 13.3.

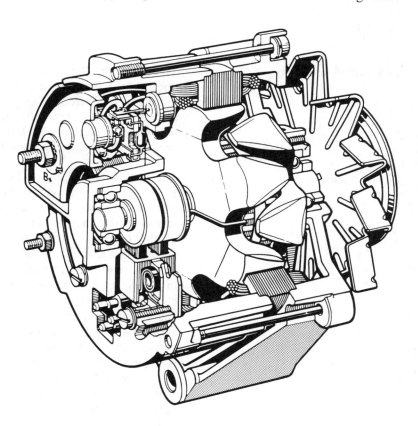

Fig. 13.3

13.4 Generator output voltage control

An explanation of how a variable resistor is used to control the field current, and hence the generator output voltage, is given in Chapter 12. In the charging system (Fig. 13.4), it is the regulator contacts opening and closing that control the size of the field current. A resistor is connected across the contacts and is switched into the field circuit whenever the contacts open. This reduces the field current. The average value of field current, and hence output voltage, is therefore related to the

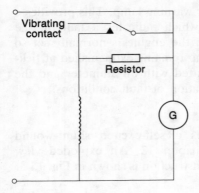

Fig. 13.4

frequency of contact vibration. When the contacts open, a high e.m.f. is induced into the field windings as a result of the flux in the magnetic field collapsing through the turns of the winding, and an induced current flows. The resistor provides a path for this current and prevents excessive arcing across the contacts.

13.5 Battery charging

To ensure that the battery is kept in a high state of charge, it is important that the generator output is used economically. The generator is controlled not only over changes in speed but also according to the state of charge of the battery and the loads placed upon the battery. Two widely used methods of control are **constant-voltage** and **constant-current** control.

Constant-voltage control

The terminal voltage of a lead acid battery varies from about 1.8 V per cell when fully discharged to about 2.5 V per cell when fully charged. By keeping the generator output voltage constant, the charging current varies with the state of charge of the battery. The battery voltage opposes the generator voltage and the size of the charging current is proportional to the difference between these two voltages. When the battery is in a low state of charge, this difference is large and a large charging current passes through the battery. As the battery charges up, the current reduces as the difference between generator and battery voltage decreases. This type of constant-voltage control therefore ensures a high charge current from the generator when the battery is discharged, and prevents overcharging by reducing current as the battery charges up. However, it has the disadvantage in that the rapid rise of battery terminal voltage during the initial stages of charging reduces the generator output too quickly. The ideal situation is for the charging current to stay at a maximum safe value until charging is almost complete, then tapering off to a **trickle charge**.

Constant-current control

When constant-current control is used, the generator output voltage increases as the battery voltage increases. The difference between the two voltages remains constant and therefore the charge current remains constant.

The disadvantages of both constant-voltage and constant-current control can be overcome by combining some of the features of both methods into one system known as current–voltage control.

13.6 Current–voltage control

A typical circuit for current–voltage control in a charging system is shown in Fig. 13.5. The battery is initially charged at a high rate which is controlled by the current regulator. When the charging circuit voltage reaches a predetermined value, the voltage regulator operates and reduces the charge current to a trickle charge.

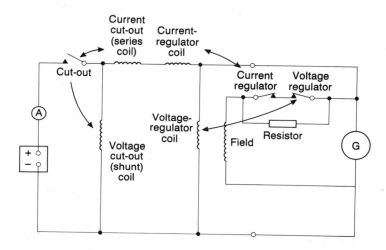

Fig. 13.5

The cut-out operates by the combined effect of both the shunt and series coils. The voltage and current regulators are operated respectively by the shunt and series coils. The shunt coil is sometimes termed the voltage coil. The series coil is sometimes termed the current coil.

Cut-out operation

The shunt and series coils are both wound on the cut-out relay core in the same direction as each other. That is to say, when the generator produces an output voltage, current passes through both coils and each produces flux in the same direction. The resulting magnetic flux in the relay core is proportional to the sum of the ampere turns of the shunt and series coils. When the cut-out contacts close the magnetic field, developed by the series coil, increases due to the charge current passing from the generator to the battery. The series coil has fewer turns than the shunt coil and, although it assists the shunt coil in closing the cut-out contacts, its contribution to the resultant flux is very small. As the current in the shunt coil is in the same direction as the charge current the flux in the relay coil is greatly increased. This ensures that the cut-out contacts stay firmly closed.

When the generator output voltage is low, the flux produced is insufficient to close the cut-out contacts and the battery remains disconnected from the generator. When the generator voltage falls below the battery voltage, a reverse current flows through the series coil. This produces a magnetic flux which opposes the existing field and accelerates the opening of the cut-out contacts. The reverse current is essential for correct cut-out operation. If the contacts remain closed, the battery will discharge through the armature of the generator and attempt to drive it as a motor. The armature is held fast by the engine drive belt and it would overheat and eventually be destroyed.

Current-regulator operation

The current-regulator coil is connected in series with the generator and is energised by current passing from the generator to the battery. When the current through the coil reaches a maximum safe value, the regulator is designed so that the contacts open. A resistor is then connected into the field circuit and this reduces the generator output voltage, and hence the output current. As the current decreases, the contacts close again disconnecting the resistor from the field circuit. The current is limited between a predetermined maximum and minimum value by the contacts continuously opening and closing.

Voltage-regulator operation

The voltage-regulator coil is connected across the generator and the battery terminals. When the line voltage reaches a predetermined value, the regulator energises and the contacts open. A resistor is connected into the field circuit and the regulator operates similarly to the current regulator. The frequency of operation of the contacts increases as the battery voltage increases and maintains the generator output voltage within safe limits. When the voltage regulator operates, the current from the generator is at a sufficiently low level for the current-regulator contacts to remain closed.

13.7 Regulator relays

The shunt and series relay coils are connected in the charging system in the same way as a voltmeter and ammeter are connected in circuits, i.e. in parallel and series, respectively. The resistance of the shunt coil is required to be high to prevent it from loading the source and, in the case of the series coil, the resistance is required to be low so that current in the circuit is not reduced. Shunt coils therefore have many turns of copper wire, whereas series coils comprise only a few turns of heavy-gauge wire. Changes in coil resistance will cause the current through the coils to change and the relays will no longer operate at the intended level of the generator output.

The resistance of copper increases with increase of temperature. This is not so important in the case of series coils because the change in resistance for these heavy-gauge wire coils is so small that it can be ignored. The cut-out and voltage-regulator shunt coils, however, have many turns of wire and their resistance will vary greatly with changes of temperature. To compensate for temperature changes many control units use **swamp resistors** to limit the resistance changes and **bimetallic strips** to vary the spring tension applied to the armature of the relays.

The bimetallic strip (Fig. 13.6)

Two strips of different metals are welded together to form a bimetallic strip. Because the two metals expand by different amounts when heated, the strip bends as temperature rises and resumes its original shape as temperature falls. When a bimetallic strip is mounted behind the relay armature control spring, increases in temperature cause the spring tension on the armature to be reduced and decreases in temperature cause it to be increased. This compensates for changes in the relay energising current.

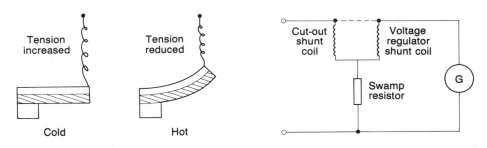

Fig. 13.6 The bimetallic strip   Fig. 13.7

The swamp resistor (Fig. 13.7)

A swamp resistor has a larger resistance than do shunt coils and is made of an alloy with a low **temperature coefficient of resistance**. By connecting it in series with the shunt coils, changes in coil resistance due to temperature changes have little effect on current flow through the winding, as the coil resistance is only a small part of the total circuit resistance.

Temperature coefficient of resistance

Increase of temperature causes an increase of resistance in all pure metals. The resistance change is proportional to the change of temperature over the general working temperature range of most equipment. The ratio of the increase of resistance per °C rise in temperature to the resistance at 0°C is called the temperature coefficient of resistance of the metal.

In the case of insulating materials and substances such as carbon and electrolytes, an increase in temperature causes a decrease in resistance and these materials are said to possess a negative temperature coefficient of resistance.

14 The ancillaries sub-system

14.1 Introduction

A variety of electrical components comprise the ancillaries sub-system. They include windscreen wiper motor, horn, direction indicating lights and stop lights as well as other components which monitor the engine performance, such as the oil-pressure warning light, coolant temperature gauge and fuel-level gauge.

A typical circuit diagram of the system is shown in Fig. 14.1. All the separate circuits, except for the horn, are connected to the battery through the ignition switch and are sometimes called the ignition ancillary circuits.

14.2 Stop lights

Stop lights are controlled by a stop light switch which is operated by the brake pedal. The two most common types of switches used are shown in Fig. 14.2, one being operated by a mechanical linkage connected to the brake pedal and the other by hydraulic pressure produced in the brake system. When the brake pedal is depressed, the contacts close and the stop light circuit is completed.

14.3 The horn

Most modern horns are designed to produce a sound consisting of a note of **high pitch** which is superimposed on a note of **low pitch**. Horns operate on an **electromagnetic vibrator** principle, as illustrated in Fig. 14.3.

When the switch is closed, current flows and the magnetic field produced in the core causes the armature to be attracted to the core. The contacts open, the coil de-energises and the armature is pulled back by spring action, thereby closing the contacts again. This process repeats itself for the time that the switch is closed at a frequency that is largely dependent on the strength of the spring. A vehicle horn has a **flexible diaphragm** and a **tone disc** attached to the armature of an electromagnetic vibrator. When the horn switch is pressed, the vibrator

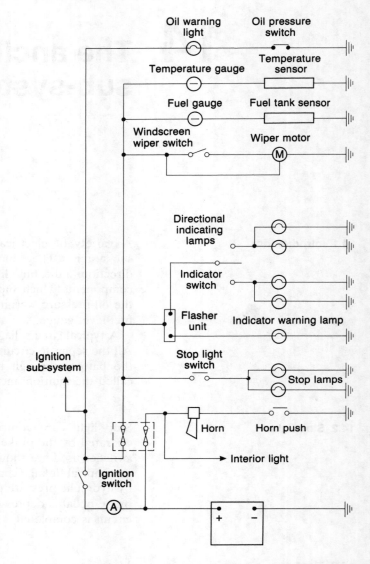

Fig. 14.1

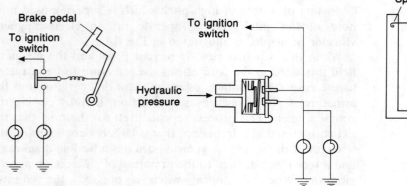

Fig. 14.2 Stop light switches

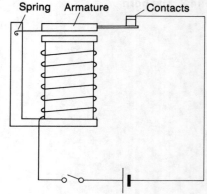

Fig. 14.3

operates and the armature moves backwards and forwards. The diaphragm produces the low-pitch note and the tone disc the high-pitch note.

14.4 Direction indicators

These are controlled by a relay or **flasher unit** which switches the lights on and off in a continuous cycle.

When the ignition switch is closed and the indicator switch is operated for a right or left turning signal, current flows through the **actuating wire**, the ballast resistor, the relay coil and lamp filaments. The lamps are not lit at this stage because the ballast resistor limits the current in the circuit. The actuating wire heats up and increases in length. This causes the main contacts of the relay to close under spring tension and since the resistance of the relay coil is small compared with the ballast resistor, current now very largely bypasses the actuating wire and the ballast resistor and the lamps are lit. The increased current in the coil produces a strong magnetic field which maintains the main contacts closed and closes the secondary contacts. These connect the indicator warning lamp to the battery which lights and gives a visual indication to the driver.

The actuating wire cools and contracts and the main contacts are forced open. The indicator lamps are extinguished and because a smaller current now flows through the coil again, the secondary contacts also open to extinguish the warning lamp. This process repeats itself at a rate of between 60 and 120 cycles per minute.

Failure of one of the indicator lamps causes only half of the normal current to flow through the coil. The magnetic effect is

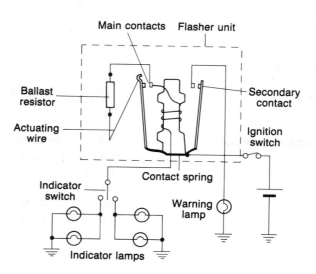

Fig. 14.4

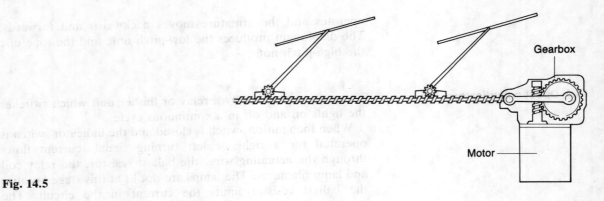

Fig. 14.5

reduced and the main contacts open more quickly. The frequency of operation of the circuit is increased. The secondary contacts operate at the increased rate or may fail to close. A visual fault indication is given to the driver by the increased rate of operation of the warning lamp or its failure to light.

14.5 Windscreen wipers

Rubber wiper blades are linked to a small d.c. motor through a gearbox. The gearbox changes the rotary motion of the motor to a side-to-side motion required to move the blades across the windscreen (Fig. 14.5). The action of d.c. motors is explained in Chapter 15.

A **limit switch** is connected in parallel with the windscreen wiper motor switch and is mechanically linked to the gearbox. When the motor control switch is operated, the mechanical link from the gearbox causes the limit switch to close. This remains closed, even when the motor control switch is opened, and current to the motor is maintained through the limit switch, causing the wiper blades to continue their movement. The gearbox motor continues to drive until a certain position is reached, thus opening the limit switch. Current flow to the motor is interrupted and the blades come to rest in their **parked** position. A typical circuit is shown in Fig. 14.6.

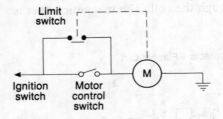

Fig. 14.6

14.6 Data transmission

Engine temperature, oil pressure and the fuel level are monitored using data-transmission systems. The physical quantities of heat, pressure and fuel level are converted to an electrical signal by a **transducer**. This signal is passed along the **data link** to a **receiver** where it is converted to some form of display acceptable to the human senses.

Engine temperature

The transducer commonly used for monitoring the engine temperature is a **thermistor**. This is a device which has a negative temperature coefficient of resistance, whereby its resistance falls significantly with increases in temperature. This change of resistance produces changes in current flow proportional to the temperature changes. The receiver is a milliammeter which is calibrated to give a visual display of the engine temperature.

Oil pressure

A pressure-operated switch with normally closed contacts is screwed into the engine block. Before the engine is started, or whenever the pressure is low, a warning lamp on the instrument panel is illuminated. When the engine is running, and provided that the oil pressure reaches a satisfactory value, the switch contacts are forced open and the oil warning light goes out.

Fuel level

The transducer is usually a variable resistor fitted inside the fuel tank and which is controlled by a float. This float moves up and down dependent upon the fuel level and causes a brush contact to change its position along the variable resistor, so changing the current flow in the data link. The receiver, or fuel gauge, is usually a **moving-iron** type instrument.

A small iron vane carrying a pointer is mounted on a pivot whose position is determined by the magnetic effect produced when current flows through a winding. When this type of instrument is used as a fuel gauge it has two windings known as the **control coil** and the **deflection coil**. The meter is designed so that the deflection of the pointer is proportional to the current in the deflection coil. Figure 14.7 shows the action of the circuit when the fuel tank is empty and when it is full.

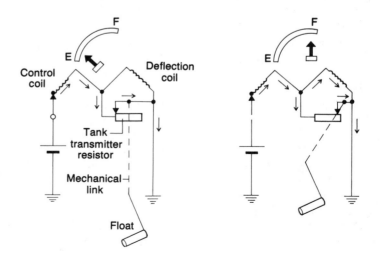

Fig. 14.7

The float is at its lowest position when the tank is empty and the transmitter resistance is at a minimum. The deflection coil is shunted by the resistor and current flow is through the control coil, as shown in Fig. 14.7. The moving iron aligns itself with the magnetic field produced by this coil winding and the pointer indicates an empty tank.

The transmitter resistance increases as the fuel tank is filled until it reaches a maximum, when the magnetic field is modified by the magnetic field produced by the deflection coil. The pointer now indicates full.

15 d.c. motors

15.1 d.c. motor principle

An electric motor converts electrical energy to mechanical energy. Its component parts are similar to those of a generator. Both have an armature, field windings and brush gear. The field windings of a motor can also be series wound, shunt wound, compound wound or separately excited, as were the generator field windings.

The operation of a motor relies on the force exerted on a current carrying-conductor situated in a magnetic field. When a d.c. supply is connected to the armature current flows through the armature winding, producing a magnetic field around it. The reaction of the armature field with the main field produces a force which acts on the armature conductors. This force causes the armature to rotate, thereby rotating the load coupled to the armature shaft.

The basic configurations for d.c. motors are shown in Fig. 15.1.

15.2 Generated e.m.f. and terminal voltage

As with a d.c. generator, an e.m.f. E_g is induced in the armature. This is explained in detail in Chapter 12.

However, in a d.c. motor, this e.m.f. is a **back e.m.f.** and opposes the applied e.m.f. V. A voltage drop (I_aR_a) occurs within the motor due to the combined resistance of the armature winding, commutator and brushes R_a and the armature current I_a. Figure 15.2 illustrates this action and by applying Kirchhoff's voltage law

$$E_g = V - I_aR_a$$

or, by rearrangement,

$$V = E_g + I_aR_a.$$

Since V exceeds E_g, the machine operates as a motor.

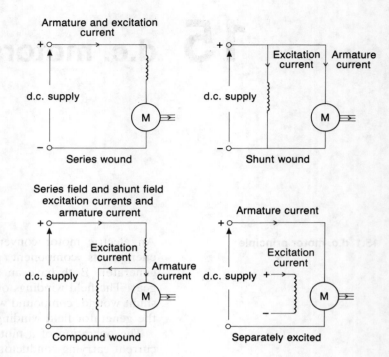

Fig. 15.1

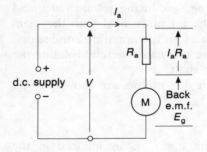

Fig. 15.2

Armature current

In general, R_a is very small and the voltage drop, $I_a R_a$, is small. Consequently, the difference between V and E_g is also small.

Example 15.1

225 V is applied to the armature of a d.c. motor which has an armature resistance of 0.1 Ω. What is the generated back e.m.f. if the armature current is 75 A?

$$E_g = V - I_a R_a.$$

The voltage drop due to the armature resistance is

$$I_a R_a = 75 \times 0.1 = 7.5 \text{ V}.$$

Therefore,

$$E_g = 225 - 7.5 \text{ V}$$

$$E_g = 217.5 \text{ V}.$$

By transposing the equation $E_g = V - I_a R_a$

$$I_a = \frac{V - E_g}{R_a}.$$

The applied voltage is generally maintained constant and changes in the back e.m.f., caused by small changes in motor speed, lead to large changes in armature current.

Example 15.2

Using the values for the motor in Example 15.1, what is the percentage increase in armature current if an increase in load reduces the motor speed by 5%?

A reduction in motor speed of 5% reduces the back e.m.f. by 5%. Therefore

$$E_g \text{ (at lower speed)} = \frac{217.5}{100} \times 95$$

$$= 206.6 \text{ V.}$$

$$I_a = \frac{V - E_g}{R_a}.$$

$$I_a \text{ (at lower speed)} = \frac{225 - 206.6}{0.1}$$

$$= 184 \text{ A.}$$

The change in $I_a = 184 - 75 = 109$ A.

The percentage increase in $I_a = \frac{109}{75} \times 100$

$$= 145.3\%.$$

15.3 Power conversion

When operating off-load, a motor accelerates until the back e.m.f. almost equals the applied voltage. A condition of equilibrium is reached when the electrical power drawn from the d.c. power source is equal to the losses incurred in driving the motor at a particular speed.

The speed of the motor decreases when it is loaded, resulting in a fall in the value of the back e.m.f.

This reduction in speed also produces large increases in armature current and therefore the electrical power drawn from the d.c. supply is greatly increased. A new condition of equilibrium is reached at a slightly lower speed where the power drawn from the supply is enough to drive the motor and its load at the new speed.

Changes in load produce changes in the armature current and therefore changes in the electrical power drawn from the d.c. supply.

15.4 Speed control

Since

$$I_a = \frac{V - E_g}{R_a}$$

then changes in either the applied voltage V or the back e.m.f. E_g cause changes in armature current I_a, since R_a is constant.

Because the torque produced by a motor is proportional to armature current, motor speed will increase or decrease accordingly. By controlling the applied voltage, or the field-excitation current, and therefore the back e.m.f., then the speed of a motor can be controlled.

Varying the applied voltage

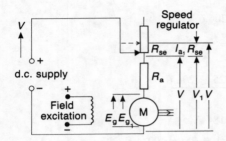

Fig. 15.3

A speed regulator in the form of a variable resistor is connected into the armature circuit. Figure 15.3 shows the circuit with the regulator adjusted so that no additional resistance is added and the motor is running at its normal speed. At this speed

$$V = E_g + I_a R_a.$$

By varying the regulator to increase the total resistance in the armature circuit to $(R_a + R_{se})$, the voltage applied to the armature is reduced and armature current falls. The motor speed falls and the back e.m.f. falls. The speed reduces to a new steady speed when

$$V = E_{g_1} + I_{a_1}(R_a + R_{se})$$

and since $V = V_1 + I_{a_1}R_{se}$, the change in terminal voltage is

$$V - V_1 = I_{a_1}R_{se}.$$

The back e.m.f. is proportional to the motor speed and the strength of the magnetic flux which, in this case, is constant.

$$E_g \propto N \times \Phi,$$

where N is the speed of the motor in rev/min and Φ is the magnetic flux.

Because the voltage drop $I_a R_a$ is usually small, then

$$V \propto N \times \Phi \text{ (approximately)}$$

therefore,

$$N \propto \frac{V}{\Phi}.$$

This shows that the motor speed is approximately proportional to the applied voltage in a shunt-wound or separately wound machine. Thus, by varying the armature voltage, the speed of the motor can be controlled from zero up to the normal running speed.

This form of speed control is used frequently to start large motors, thereby limiting the armature current during the starting period. It has three main disadvantages, as follows:

1 power is wasted in the resistor $(I_a^2 R_{se})$;
2 the resistor is required to be physically large and is usually expensive; and

Varying the field current

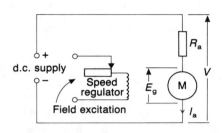

Fig. 15.4

3 starting torque is reduced and changes in load cause considerable speed changes.

In this case, a speed regulator is connected into the field circuit. Figure 15.4 shows the circuit with the regulator adjusted so that no additional resistance is added and the motor is running at its normal speed. At this speed $V = E_g + I_a R_a$.

When the regulator is adjusted to increase the field resistance, the field current reduces and the magnetic field reduces, causing a corresponding reduction in the generated back e.m.f.

Since $V = E_g + I_a R_a$ when V and R_a are constant, then if E_g decreases, I_a increases. Motor speed increases, causing an increase in E_g which in turn reduces I_a back towards its original value. A new speed is achieved when

$$V = E_{g_2} + I_{a_2} R_a,$$

where E_{g_2} and I_{a_2} represent the values of back e.m.f. and armature current at this new speed. Because this armature current tends towards its original value then the relationship between motor speed and the strength of the magnetic field is

$$E_g \propto N \times \Phi.$$

Therefore,

$$N \propto \frac{E_g}{\Phi} \, .$$

The speed is therefore inversely proportional to the strength of the magnetic field. Varying the field current provides speed control from normal speed up to a maximum speed determined by the losses and the load.

This form of speed control is widely used and, in shunt and compound motors, a field regulator is connected in series with the field winding. In series motors, it is connected in parallel with the field winding.

In the case of shunt and compound machines, maximum speed is achieved when a maximum resistance is added. For series machines, a maximum speed is attained with minimum resistance added. This is illustrated in Fig. 15.5.

15.5 Load characteristics

The manner in which the speed and the torque of motors vary as the load on the machine changes is extremely important and must be considered when selecting a motor for a particular function. The characteristics of the four main types of motor are as follows.

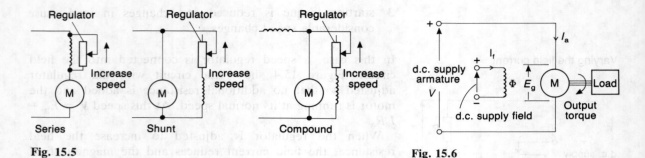

Fig. 15.5

Fig. 15.6

Separately excited motor (Fig. 15.6)

Changes in the load do not affect the field current I_f and, consequently, the magnetic flux Φ remains constant. Therefore, the motor torque is directly proportional to the armature current I_a.

The applied voltage and flux are constant and only a small reduction in the back e.m.f. E_g is required to allow the full load armature current to flow. As the load increases, the motor speed stays approximately constant (Fig. 15.7).

This type of motor is very flexible in its operation and is easily controlled, because the field circuit is independent of the armature. However, due to the cost of installing a separate d.c. supply, such machines are not widely used.

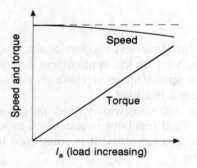

Fig. 15.7 Separately excited motor – speed and torque characteristics

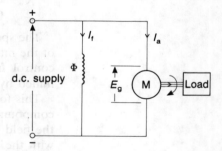

Fig. 15.8

Shunt motor (Fig. 15.8)

The field current I_f is virtually constant over the normal load range of the motor. The speed and torque characteristics, shown in Fig. 15.9, are similar to those of the separately excited motor.

The shunt motor is often used to drive lathes, drills and other light machine tools and is generally considered to be a

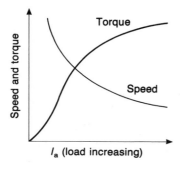

Fig. 15.9 Shunt motor – speed and torque characteristics

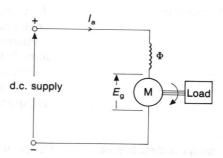

Fig. 15.10

constant-speed machine. Its starting torque is small and the shunt motor is usually started off-load or under light load conditions.

Series motor

Figure 15.10 shows that the armature current flows through the field winding. The magnetic flux produced is therefore proportional to the armature current until the pole shoes reach magnetic saturation. The torque is proportional to the armature current and the flux and, for a series motor, is therefore proportional to I_a^2. As the load increases the torque increases rapidly until saturation is reached.

Because speed is inversely proportional to the flux, it decreases quickly as the load increases (Fig. 15.11).

A series motor is used where a high starting torque is required, such as with the engine starter motor. Under no load conditions, a series motor accelerates rapidly and damage will occur. It is therefore very rare for a series motor to be used without a permanently coupled load.

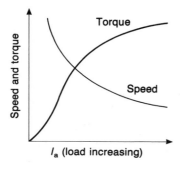

Fig. 15.11 Series motor – speed and torque characteristics

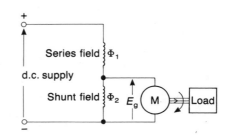

Fig. 15.12

Compound motor (Fig. 15.12)

The compound motor combines the characteristics of both the series and shunt motors and can provide almost any required characteristic depending on the strength of the series and shunt fields.

They are generally used for driving pumps and compressors and other heavy-duty machinery where a large starting torque and speed control are required.

15.6 Reversing the direction of rotation

Reversal of the direction of rotation of a motor with field windings is achieved by reversing the connections of either the armature or the field windings. In permanent magnet motors, the direction of the field cannot be changed so the armature current must be reversed to change the direction of rotation.

16 d.c. machines – further considerations

16.1 Efficiency of d.c. machines

The efficiency of a machine is the ratio of the power output to the power input and is normally expressed as a percentage.

$$\text{Efficiency} = \frac{\text{output power}}{\text{input power}} \times 100\%,$$

or

$$\text{Efficiency} = \frac{\text{output power}}{\text{output power} + \text{losses}} \times 100\%.$$

Large d.c. machines are generally more efficient than small machines because the mechanical losses do not increase with machine size at the same rate as the output power. The efficiency of motors and generators varies between 40% for low horsepower machines to over 90% for machines of around 10 h.p. (7.5 kW) and above, although the efficiency of any machine varies as operating conditions change.

Horsepower is still widely used to represent the power rating of a machine. Its relationship with the SI unit of power is

1 horsepower = 746 watts.

Example 16.1

A motor draws 20 A from a 110 V d.c. supply. What is the efficiency of the machine if it produces an output power of 2 kW?

$$\text{Output power} = 2000 \text{ W}$$

$$\begin{aligned}
\text{Input power} &= V \times I \\
&= 110 \times 20 \\
&= 2200 \text{ W}.
\end{aligned}$$

$$\begin{aligned}
\text{Efficiency} &= \frac{2000}{2200} \times 100\% \\
&= 90.9\%.
\end{aligned}$$

Example 16.2

A motor has an output power of 10 h.p. The losses of the machine are equivalent to 0.7 h.p. What is the efficiency of the motor?

$$\text{Efficiency} = \frac{\text{output power}}{\text{output power} + \text{losses}} \times 100\%$$

$$= \frac{10}{10 + 0.7} \times 100\%$$

$$= \qquad 93.5\%$$

16.2 Power losses

The power losses inherent in machines can be separated into three main areas:

1. mechanical losses due to **friction** and **windage**;
2. iron **losses** due mainly to the magnetic properties of the armature core; and
3. electrical or **copper losses** as a result of passing current through the armature and field windings.

The iron and copper losses represent the main portion of wasted power.

Iron losses

Iron losses are classified as being due to **eddy current** loss and **hysteresis** loss.

1. Because iron and steel are conductors, when an armature rotates in the magnetic field, e.m.f.s are induced into the core. For a solid core, these produce circulating, or eddy currents around the core parallel to the armature shaft. Eddy currents generate heat that can damage the insulation of the armature conductors. The size of eddy currents is reduced by laminating the armature (Fig. 16.1). Laminations are usually thin steel plates separated by an insulating

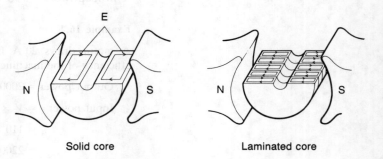

Fig. 16.1   Solid core   Laminated core

material which effectively increase the electrical resistance of the core due to

$$R = \frac{\varrho l}{a}$$

and the area a being decreased.

2   The rotation of an armature between alternate north and south poles of a magnet causes the armature to undergo rapid reversals of magnetism. Whenever a material is magnetised in a particular direction, it retains some residual magnetism which must be eliminated before the material can be magnetised in the opposite sense. In d.c. machines, the elimination of the residual magnetism at each reversal wastes energy in the form of heat being dissipated in the core. This is called hysteresis loss. Some grades of alloy steel produce small hysteresis loss and these are used in the manufacture of armature cores.

Copper losses

These result from the resistance of the armature windings and the field. The contact resistance between brushes and the commutator are also usually considered to be part of the copper losses.

16.3 Armature reaction

Armature reaction, or **cross magnetisation**, is the interaction between the magnetic field due to armature current and the external field. The result is a combined field in which the external field lines of flux are distorted from their original direction. This is illustrated in Fig. 16.2 for a generator and in Fig. 16.3 for a motor, where, due to the relative polarities of the interacting fields, the resultant field is distorted in the direction of rotation for a generator and against the direction of rotation for a motor.

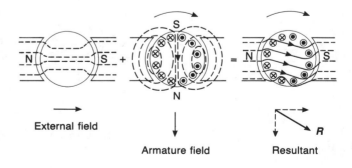

Fig. 16.2 Generator cross magnetisation

External field + Armature field = Resultant field

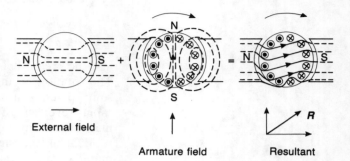

Fig. 16.3 Motor cross magnetisation

During commutation, each coil is short-circuited in turn to the brushes. Because of armature reaction, the e.m.f. induced into other coils is sensed by the brushes. When an e.m.f. is induced into the short-circuited coil, **sparking** occurs between the brushes and the commutator segment connected to the coil as the armature rotates and the short circuit is removed. This burns the brushes and damages the commutator, leading to a poor electrical contact between them. Sparking also makes the output of a d.c. generator less smooth and is a source of radio interference.

The sparking can be reduced by moving the brushes into a **neutral plane** as shown in Fig. 16.4, either forwards in the direction of rotation for a generator or backwards for a motor. However, this is not very satisfactory as changes in load conditions cause changes in armature current and, consequently, changes in position of the neutral plane.

Armature reaction is effectively neutralised by positioning small electromagnetic poles known as **interpoles** or **commutating poles** midway between the main poles with their coils connected in series with the armature. These interpoles produce a magnetic field whose strength is proportional to the

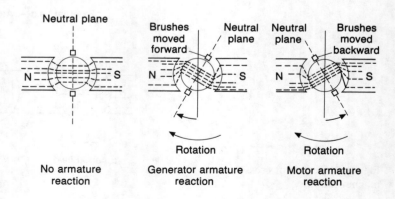

Fig. 16.4

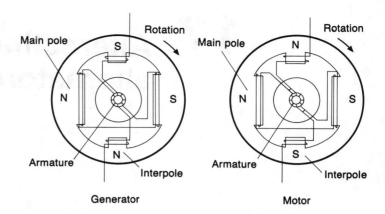

Fig. 16.5

armature current, thereby neutralising the armature fields for all load conditions. Interpoles are connected to produce the sequence of magnetic polarities shown in Fig. 16.5 in order to eliminate the effects of armature reaction. For generators, an interpole is positioned to come before a main pole of the same polarity for the direction of rotation, and to follow a main pole of the same polarity for motors.

17 Alternating current and alternators

17.1 Introduction

Most electrical power **mains** are alternating current supplies. It is fairly easy to increase or decrease alternating voltage without appreciable power losses through the use of transformers, whereas direct voltages produce significant power losses when changed.

Example 17.1

A d.c. generator has a terminal voltage of 400 V and supplies 150 A to a load situated at some distance from the generator. The total resistance of the connecting cable is 1 Ω. Determine the power loss and voltage drop in the cable (Fig. 17.1).

$$\text{The power loss in the cable} = I^2 R$$
$$= (150)^2 \times 1$$
$$= 22\ 500 \text{ W}.$$

$$\text{The voltage drop in the cable} = I \times R$$
$$= 150 \times 1$$
$$= 150 \text{ V}.$$

A 400 V alternator is used to supply the same load with a 10 to 1 step-up transformer at the supply end of the cable and a 10 to 1 step-down transformer at the load end. The transformers are assumed to be lossless. Determine the new values of the power loss and the voltage drop.

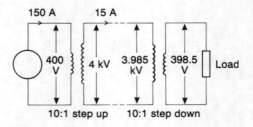

Fig. 17.1

The 10 to 1 step-up transformer converts the alternator output to 4000 V at 15 A.

The power loss in the cable $= I^2R$

$$= (15)^2 \times 1$$

$$= 225 \text{ W.}$$

The voltage drop in the cable $= I \times R$

$$= 15 \times 1$$

$$= 15 \text{ V.}$$

The available voltage at the end of the cable is

$$4000 - 15 = 3985 \text{ V.}$$

This is stepped down by the transformer to 398.5 V.

Using a.c. for the distribution results in a loss of 1.5 V and 225 W, whereas using d.c. results in a loss of 22 500 W and a voltage drop of 150 V.

The transmission of large amounts of electric power is best done at high voltages. Transformers step up the voltage at the power station for transmission and then step it down for use in factories and the home.

Direct current effectively flows through a conductor in one direction, determined by the polarity of the supply, whereas alternating current flows back and forth as the direction of the voltage changes. The **waveform** of a.c. mains is in the form of a sine wave.

17.2 Frequency

Sin θ

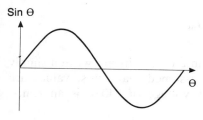

θ

Fig. 17.2

Figure 17.2 shows one complete **cycle** of a waveform for a sine wave, i.e. it rises from zero to a maximum positive value, then falls back through zero to a maximum negative value and finally rises back to zero, where a positive value indicates a current flow in one direction and a negative value the flow of current in the opposite direction. The number of these cycles that occur in one second is called the **frequency** and is measured in hertz (Hz). The normal mains frequency in the home is 50 Hz, indicating that the voltage completes 50 cycles of the waveform in one second.

17.3 a.c. voltage and current values

Figure 17.3 shows the maximum, or **peak amplitude** of the sine wave measured from the zero horizontal axis. This peak value is denoted as V_{pk}, or $\hat{V}$ in the case of a voltage wave form, or I_{pk} or $\hat{I}$ in the case of a current wave form. For every complete cycle of a.c. there are two peak values, one for the positive half cycle and one for the negative half cycle. The difference between these two peak values is called the **peak-to-peak** value

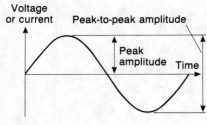

Fig. 17.3

$(V_{\text{p-p}}$ or $I_{\text{p-p}})$. This value is twice the peak value for a sine wave and is sometimes used as a way of expressing the amplitude of the wave.

a.c. and d.c. are both used to provide electric power. To provide a comparison between a.c. and d.c., suppose that an electric heater is to be used first with a d.c. supply and then with an a.c. supply and that the heating effect is to be identical in each case. Assume that the heater has a resistance of 1 kΩ and, when fully on, dissipates 1 kW. For d.c.

$$P = I^2R$$

or, by transformation,

$$I = \sqrt{\frac{P}{R}} = \sqrt{\frac{1000}{1000}} = \sqrt{1} = 1 \text{ A}.$$

When the a.c. supply is connected in place of the d.c., its voltage is adjusted until the heater again dissipates 1 kW. The alternating current flowing in the circuit is equivalent to the direct current and is said to have a **root mean square** of 1 A.

The r.m.s. current is defined as having the same value as the direct current which would give the same heating effect. For voltage and current, r.m.s. values are denoted as V_{rms} or I_{rms} to distinguish them from peak values. The relationship between r.m.s. and peak values is

$$V_{\text{rms}} = \frac{V_{\text{pk}}}{\sqrt{2}} = 0.707 \times V_{\text{pk}}$$

or

$$I_{\text{rms}} = \frac{I_{\text{pk}}}{\sqrt{2}} = 0.707 \times I_{\text{pk}}.$$

When voltages or currents in a.c. circuits are written simply as 240 V or 10 A a.c., it is assumed that r.m.s. values are implied. The domestic mains voltage of 240 V is an r.m.s. value.

17.4 Generating alternating current

Chapter 12 explains how an a.c. sinusoidal voltage is induced in a coil which rotates in a magnetic field. When the ends of the coil are connected to slip rings, the induced a.c. voltage can be applied to an external circuit and the system forms a simple a.c. alternator.

It is usual in d.c. generators for the coil to rotate and the field to be stationary. In alternators the field winding is wound either on the rotating element, called the **rotor**, or on the stationary element, known as the **stator**. When being run as an alternator, it does not matter whether a rotating winding cuts a stationary field or a rotating field cuts a stationary winding.

Rotating-coil alternator

Some small alternators are usually of this type and are similar to d.c. generators in their operation. Field coils are wound round one or more pairs of laminated pole shoes fixed to the stator. The coils are connected to a d.c. supply so that they produce alternate north and south magnetic poles. The armature comprises a number of coils connected in series which terminate at slip rings. The induced voltage in the armature is connected through brushes to the load.

Rotating-field alternator

Large alternators are usually rotating-field alternators which have a stationary coil, allowing the generated voltage to be connected directly to the load. The advantage of this system is that there is no requirement to provide slip rings and brushes capable of withstanding high voltages and conducting high values of current. Permanent magnets are sometimes used to produce the field but, more usually, field windings are employed. In this case only low values of voltage and current, required to excite the field, affect the slip rings and brushes.

The stator comprises a laminated steel core in which the armature coils are embedded. Rotor design is dependent to some extent on the speed at which it is to be driven. At high speeds, large forces act on the rotor and its winding. To cope with these forces, high-speed machines (typically above 1200 rev/min) use small cylindrical rotors which have the field windings fixed into slots cut into the rotor. Slower running machines often use a **salient-pole** rotor which has a number of separately wound pole shoes bolted to it. The field coils are connected in series, or in series–parallel groups, and produce two or more distinct magnetic poles. A d.c. supply is connected to the complete winding through slip rings and brushes.

17.5 Field excitation

Field windings need to be energised by a direct current. Some alternators use a portion of the a.c. output which is **rectified** and used to excite the field windings. A **rectifier** is a circuit which converts a.c. to d.c. and is explained in detail in Chapter 20.

These machines are self excited and rely on residual magnetism for their operation.

Separately excited machines are also used and some larger machines use a small d.c. generator called an **exciter** to provide the excitation current.

17.6 The single-phase alternator

Figure 17.4 represents a two-pole rotating-field single-phase alternator. Alternating voltages are induced in the stator windings as the rotor turns. The stator coils are connected such that the voltages induced in them are **series aiding** or **in phase** and the output is a sinusoidal alternating voltage.

A single-phase machine is generally used for low-power applications.

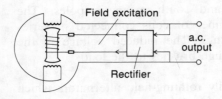

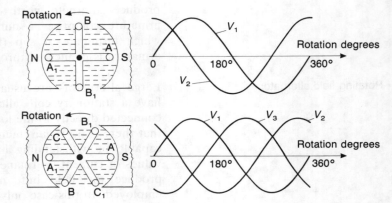

Fig. 17.4

Fig. 17.5

17.7 The polyphase alternator

A polyphase alternator is one that has two or more single-phase windings displaced around the stator or rotor. Figure 17.5 illustrates how two-phase or three-phase outputs are produced, using two or three armature windings.

Two-phase machines are seldom used but three-phase alternators are often preferred for the following reasons.

1 A larger output is possible from a three-phase machine than from a single-phase machine of the same size.
2 Motors and other machinery operate more efficiently from three-phase supplies.
3 It is more economical to transmit and distribute three-phase a.c.

The three phases are normally identified as the red, yellow and blue phases. Although it is possible to supply a separate load from each phase, it is common for phases to be connected together so that a load is common to two or three phases.

18 Reactance

18.1 Introduction

A resistor opposes the flow of current in a circuit. The amount of opposition offered is known as resistance and is measured in ohms, and remains the same value when both direct and alternating voltages are applied to a circuit containing resistance only. Inductors and capacitors also oppose the current in circuits but this opposition is known as **reactance** and its value is frequency dependent. Reactance is measured in ohms.

18.2 Inductive reactance

Inductance is a property of a coil to oppose any change in current. Inductive reactance is the actual opposition that the inductance offers to the flow of a varying current. The symbol used for reactance is X and for inductive reactance, X_L. When a current of one ampere flows under a pressure of one volt in a purely reactive circuit it is said to have a reactance of one ohm. This is similar to Ohm's law and, mathematically,

$$\frac{E}{I} = X \text{ ohms.}$$

An inductor is essentially a coil of wire, and when an ohmmeter is connected to its ends, the resistance of the wire is measured. The reactance offered by an inductor is dependent on the frequency f in hertz and the inductance L of the component in henrys. The formula used to determine X_L is

$$X_L = 2\pi f L.$$

$2\pi f$ is known as the angular frequency and is measured in radians per second. This is sometimes replaced by ω and the formula is written as

$$X_L = \omega L.$$

When the frequency is zero (direct voltage supply) X_L is zero and the only opposition to current flow in the circuit is the resistance of the wire coil in ohms.

At low alternating frequencies, f is small and X_L is small. At high frequencies when f is large, then X_L is large.

Example 18.1

A 636 µH coil is connected first to a 100 V, 100 Hz supply and then to a 100 V, 100 kHz supply. What is the reactance offered by the coil and the value of current through the coil in each case, neglecting the resistance of the coil?

a) $X_L = 2\pi f L \ \Omega$

$X_L = 2\pi \times 100 \times 636 \times 10^{-6} \ \Omega$

$X_L \simeq 0.4 \ \Omega$

$I = \dfrac{E}{X_L} = \dfrac{100}{0.4} = 250$ A.

b) $X_L = 2\pi f L \ \Omega$

$X_L = 2\pi \times 10^5 \times 636 \times 10^{-6} \ \Omega$

$X_L \simeq 400 \ \Omega$

$I = \dfrac{E}{X_L} = \dfrac{100}{400} = 0.25$ A.

18.3 Capacitive reactance

Capacitive reactance is the opposition that a capacitor offers to the flow of a varying current. The symbol used to represent capacitive reactance is X_C and is given by

$$X_C = \frac{1}{2\pi f C},$$

where f is the frequency in hertz, and C is the capacitance of the capacitor in farads.

$2\pi f$ can again be represented by

$$X_C = \frac{1}{\omega C}.$$

At zero frequency,

$$X_C = \frac{1}{0} = \infty.$$

When a direct voltage is applied to a capacitor, no direct current flows as it comprises two metal plates separated by an insulator. Since

$$X_C = \frac{1}{2\pi f C}$$

then, at low frequencies, X_C is large and at high frequencies it is small.

18.4 Current and voltage relationships

The rate of current flow in a capacitor is dependent on the rate of change of voltage. When a direct voltage is applied to a circuit containing capacitance there is no change of voltage. For a circuit which has an alternating voltage supply, the current is greatest when the voltage passes through zero and has the greatest rate of change. Figure 18.1 shows the implications of this statement in that the current is at a maximum when the voltage is zero and is said to **lead** the voltage by 90°. Conversely, when current flows through an inductor it **lags** the voltage by 90° (Fig. 18.2).

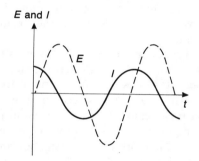

Fig. 18.1 Current through a capacitor

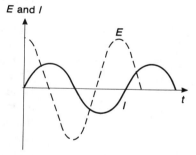

Fig. 18.2 Current through a conductor

18.5 Comparison of component properties

The opposition to current flow offered by inductors, capacitors and resistors is summarised in Table 18.1.

Table 18.1

	Inductor	Capacitor	Resistor
Opposition to d.c.	Negligible	Infinite	Ohmic value of resistor
Opposition to low frequency a.c.	Low	High	Ohmic value of resistor
Opposition to high frequency a.c.	High	Low	Ohmic value of resistor

19 The diode

19.1 Semiconductors

Insulators are poor conductors of electricity, whereas metals are good conductors of electricity. A material which is between these two extremes is called a semiconductor. The conductive nature of materials is due to their atomic structure.

Electrons orbit the nucleus of all atoms in defined **bands**. The energy level of the orbiting electrons is related to the size of the orbit, with electrons of higher energy occupying the larger outer bands. The study of semiconductors is mainly concerned with the two outer bands called the **valency** band and the **conduction** band.

Two commonly used semiconductor materials are **silicon** (Si) and **germanium** (Ge). These are both **tetravalent** materials which means that each atom has four electrons in its valency band. At **absolute zero** ($-273\,°C$) silicon and germanium adopt a crystal structure because their atoms naturally align themselves with a regular spacing between each other. Valency electrons share their orbits with neighbouring atoms and the material is held tightly together. This is called **covalent bonding**. As the temperature increases, energy is imparted to the electrons and some of them gain sufficient energy to break free from the covalent bonds and move into the conduction band. These electrons are then free to 'wander' around the crystal structure from atom to atom as with conductors and are called **free electrons**. The **conductivity** of the material is thus raised.

When an electron moves from its valency band, it leaves a space behind it called a **hole**. As this is deficient of a negative charge, it is regarded as being an equivalent positive charge. For every free electron there is a hole produced and these pairs are known as **electron–hole pairs**. Negative charges are attracted to positive charges and when a free electron meets a hole, the two **re-combine** and lose their separate identities. The production of electron–hole pairs and the rate of recombination is a function of temperature and, at room temperature, silicon and germanium are regarded as fair conductors of electricity.

Under these circumstances, the material is known as an **intrinsic** semiconductor and is of little use. Conductivity is increased to make the semiconductor more positive (**P-type**) or more negative (**N-type**) by a process known as **doping**. The material is now called an **extrinsic** semiconductor.

Doping

Doping is a process where impurity atoms are introduced to the intrinsic material. To produce N-type material, impurities such as arsenic, antimony or phosphorus are used. These are **pentavalent** materials whereby each atom possesses five valency electrons. Four of the electrons form a covalent bond with neighbouring atoms, but the fifth has no part to play and is automatically regarded as a free electron, making the material more negative. The impurity atoms are known as **donor** impurities because they 'donate' free electrons. When an impurity atom loses an electron it is regarded to effectively have a positive charge and is known as a **positive ion**.

P-type material is produced by adding **acceptor** impurities such as gallium, indium or boron which are **trivalent**. Having only three electrons in their valency bands, these atoms 'accept' an electron from a neighbouring atom which in turn leaves a hole behind it, making the material more positive. When the impurity atom accepts an electron it becomes more negative and is called a **negative ion**.

Thermally produced electrons in N-type materials blend in with the donor electrons. The thermally produced holes, however, are out of place and are called **minority carriers**. Similarly, thermally produced electrons in P-type material are also regarded as minority carriers, the **majority carriers** being holes.

19.2 The P–N junction

When P- and N-type materials are brought together, certain changes occur at their junction. The large amount of mobile holes in the P-type semiconductor and the large amount of mobile electrons in the N-type semiconductor are attracted and initially cross the junction in both directions. This is known as **diffusion**. As holes leave the P-type material, a net negative charge is left behind near the junction due to the 'uncovered' remaining negative ions. A similar net positive charge occurs near the junction in the N-type material due to uncovered positive ions and a potential difference is established across the junction. For silicon, this is in the order of 0.6 to 0.7 V and, for germanium, 0.2 to 0.4 V. Recombination takes place either side of the junction and an area which is depleted of mobile carriers is produced, known as the **depletion region** (Fig. 19.1).

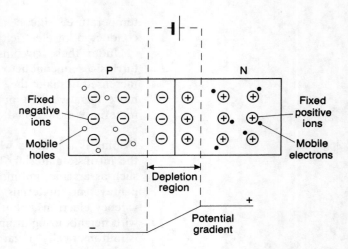

Fig. 19.1

Fig. 19.2

This P–N junction device is called a **diode**. A diode is a component which allows current to flow through it in one direction only. The symbol for a diode is shown in Fig. 19.2.

The P-type terminal is called the **anode** and the N-type terminal the **cathode**.

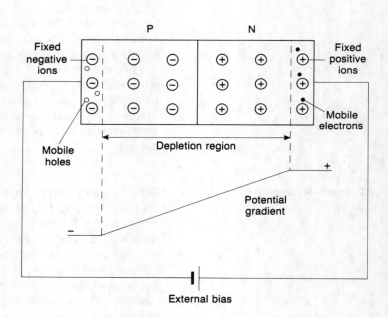

Fig. 19.3

19.3 A reverse-biased diode

When an external e.m.f. or **bias** is applied to the diode with the positive and negative terminals connected, as shown in Fig. 19.3, no current flows through the device and it is said to be reverse biased. Mobile electrons are attracted out of the cathode region to the positive terminal of the supply and mobile holes are attracted from the anode to the negative terminal of the supply. More positive and negative ions are uncovered and the depletion region is extended, thereby increasing the potential gradient across the junction.

A small leakage current flows through the device due to minority carriers. To the minority electrons in the anode and the minority holes in the cathode, the external reverse bias appears as a **forward bias** and they are attracted to each terminal of the supply, accordingly. This **small** leakage current is in the order of microamps and is temperature dependent.

19.4 A forward-biased diode

When the polarity of the bias is changed round, the junction is said to be forward biased. Provided that the external bias is large enough to overcome the potential gradient at the junction, mobile electrons flow from cathode to anode, as shown in Fig. 19.4. Conventional current therefore flows from anode to cathode. The electrons and holes readily crossing the junction recombine and this is undesirable. In practice, one side of the junction is doped more heavily than the other and because the two types of majority carriers are not produced in equal numbers then recombination is relatively small.

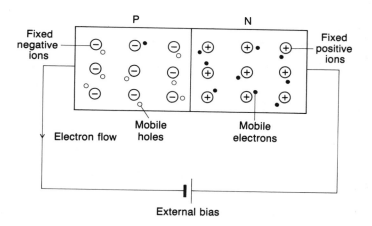

Fig. 19.4

External bias

19.5 The diode characteristic

The characteristic curve for typical silicon and germanium diodes is shown in Fig. 19.5. When the device is reverse biased, leakage current flows as explained earlier. Increasing the reverse bias causes minority carriers to be accelerated across the junction with increasing velocity and collisions occur. These collisions 'knock' further electrons free and more electron–hole pairs are produced, in turn producing more minority carriers. If the reverse bias is large enough a point known as **avalanche breakdown** is reached and this can lead to the destruction of the diode if left unchecked.

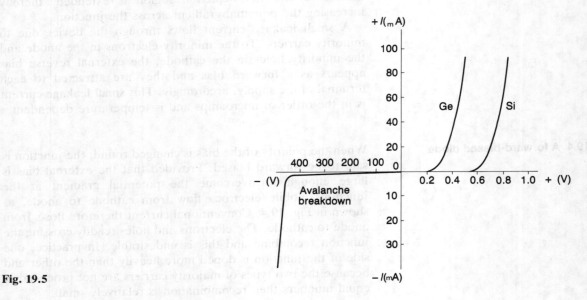

Fig. 19.5

20 Rectification

20.1 Introduction

The purpose of a rectifier is to convert a.c. to d.c. The use of such a circuit is illustrated in Chapter 17, where an a.c. alternator rectifies part of its a.c. output to provide the d.c. field-excitation voltage. The component that is essential to the operation of a semiconductor rectifier is the diode.

20.2 Half-wave rectification

A half-wave rectifier circuit is shown in Fig. 20.1 where R_L represents a load. During positive half cycles of the input voltage V_i, point A is positive and point B is negative. These polarities reverse during negative half cycles of the input voltage. During positive half cycles therefore, the anode of D_1 is positive with respect to its cathode. The diode is forward biased and current flows fron anode to cathode and through the load. A positive half cycle of voltage is developed across the load resistor. During negative half cycles, the diode is reverse biased as its anode is negative with respect to the cathode. No current flows through the diode and no output voltage is developed across the load.

The process repeats itself for every cycle of the input voltage and the output is a series of positive 'pulses'.

20.3 Full-wave rectifier

A full-wave rectifier is shown in Fig. 20.2 which uses a transformer whose secondary winding is centre tapped to earth. Two diodes, D_1 and D_2 are connected to the transformer and to the load resistor R_L, as shown. During positive half cycles of the input voltage V_i the top of the secondary winding is positive with respect to the bottom. D_1 is forward biased and D_2 is reverse biased. Current flow is through D_1, R_L and back to earth. As with the half-wave rectifier, a voltage is developed across R_L in the form of a positive half cycle.

During negative half cycles of the input voltage, the polarity

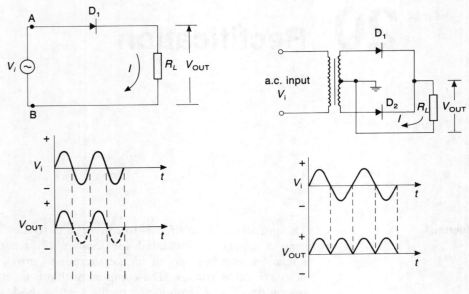

Fig. 20.1   Fig. 20.2

of the anodes of the diodes reverses. D_1 is reverse biased and D_2 is forward biased. Current flow is now through D_2, R_L and back to earth. This current flow through R_L is in the same direction as during positive half cycles of input voltage and the output voltage developed across the load resistor is in the same sense. A series of positive pulses is therefore produced at the output by both positive and negative half cycles of input.

The average d.c. output is increased to twice that of a half-wave rectifier but the circuit has the disadvantage of using a centre-tapped transformer which is a costly item. A more commonly used circuit is the bridge rectifier.

20.4 The bridge rectifier

The circuit uses four diodes connected in a bridge arrangement as shown in Fig. 20.3. During positive half cycles of the input voltage point A is positive with respect to point B. D_3 and D_4 are forward biased and D_1 and D_2 are reverse biased. Current flow is through D_3, R_L and D_4, as illustrated in Fig. 20.4.

During negative half cycles point A is negative with respect to point B. D_1 and D_2 are now forward biased and D_3 and D_4 reverse biased. Current flow is through D_2, R_L and D_1, as illustrated in Fig. 20.5.

During both half cycles of the input voltage, current flow through R_L is in the same direction and the output voltage is a series of positive pulses as with the two-diode rectifier circuit.

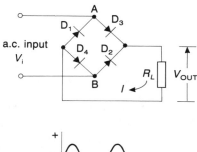

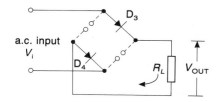

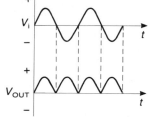

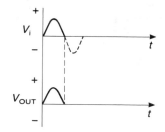

Fig. 20.3   **Fig. 20.4**

20.5 Smoothing

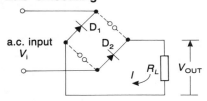

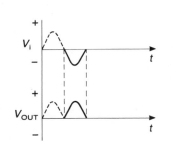

Fig. 20.5

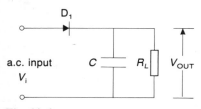

Fig. 20.6

The output voltage for both the half-wave and full-wave rectifier circuits is unidirectional but constantly changing in value. This can be **smoothed** by connecting a **reservoir capacitor** in parallel with the load resistor.

During positive half cycles of the input voltage, D_1 is forward biased and conducts. The capacitor charges rapidly to the peak value of the input voltage. The time constant for the charge is $C \times R_f$ seconds where R_f is the very low forward resistance of the diode. When the input voltage begins to fall below its peak value the voltage on the anode also begins to fall. The cathode, however, tends to remain at the peak value due to the charge on the capacitor. The anode is now negative with respect to the cathode and the diode is reverse biased and 'cuts off', that is, a high resistance value is introduced into the circuit. The capacitor discharges through the load resistor. The time constant for discharge is $C \times R_L$ seconds. This is a much longer time than the charge time and the discharge rate is consequently much slower. On the next positive half cycle the diode does not conduct until the anode voltage is greater than the potential on the cathode. The capacitor then charges to the peak value and the process repeats itself. The output waveform for the circuit in Fig. 20.6 is shown in Fig. 20.7.

When a reservoir capacitor is used to smooth the output of a full-wave rectifier, it has less time to discharge before a positive half cycle of the input causes it to charge again. This results in a smaller **amplitude ripple component**.

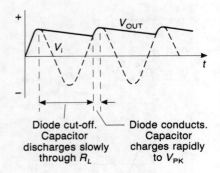

Fig. 20.7

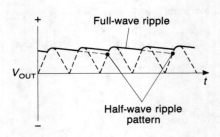

Fig. 20.8

21 Measuring instruments

21.1 Introduction

Measuring instruments are essential tools for maintaining and servicing vehicle electrical systems. General fault-finding procedures require the measurement of electrical quantities, such as voltage, current and resistance. A wide range of meters are commercially available, the most common one being the **multimeter**. As the name multimeter implies, this essentially incorporates a voltmeter, ammeter and ohmmeter into one instrument.

Historically, the most common type of multimeter used a pointer which moved over a graduated scale known as an analogue meter. These are still available but technological advances have made the digital meter more popular, where a number is displayed to denote the quantity being measured. Although digital instruments are generally no more expensive than their analogue counterparts, their extensive use is justified by the following advantages:

1 A better accuracy and sensitivity is possible.
2 User skills are reduced and therefore less reading errors occur.
3 Results are transmitted more simply.
4 Readings can be taken more quickly.

The method in which both analogue and digital meters are connected in circuits for measuring electrical quantities is described in Chapter 1. These methods are summarised as follows.

Voltage measurement

The meter leads are connected across, or in parallel with, the component(s) whose voltage is to be measured.

Current measurement

The meter is connected in series with the supply and components of the circuit. This invariably requires the circuit to be 'broken' so that the meter can be inserted in the 'break' and therefore restore continuity.

Resistance and continuity measurement

The meter leads are connected in the same manner as for voltage checks, i.e. in parallel. However, no current must be flowing when resistance or continuity checks are made or else damage to the instrument may occur. The circuit supply must be switched off.

21.2 Analogue meters

All analogue instruments depend on three torques for their operation.

Deflecting torque

The deflecting torque depends on the electrical effect to be measured for its value and causes the meter movement to be deflected from its rest position.

Control torque

This acts in the opposite sense to the deflecting torque and the meter movement adopts an equilibrium position when both torques are equal in magnitude but opposite in sense. The control torque is usually produced using springs or by gravity.

Damping torque

The damping torque ensures that the meter movement reaches its equilibrium position quickly and without undue oscillation. Three main ways of producing damping are by using **eddy currents**, **air friction** or **fluid friction**.

Analogue meters are usually **moving-coil** instruments or **moving-iron** instruments.

Moving-coil type instruments (Fig. 21.1)

A coil wound on a shaped former is positioned between the poles of a magnet. This coil is supported by pivots mounted between jewelled bearings and a pointer is attached to the coil. When current passes through the coil, a force is produced to cause it to move. This is the same as the motor operating principle. The amount of movement of the coil, and hence of the pointer, is proportional to the current flowing in the coil.

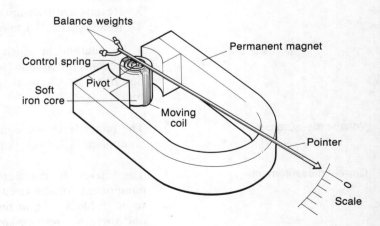

Fig. 21.1 Moving-coil type instrument

The scale over which the pointer moves is therefore **linear**. Control springs fitted to each end of the coil are often used to provide the controlling torque. When the coil moves, an e.m.f. is induced into it producing eddy currents. These oppose coil movement and provide the damping torque.

The instrument only measures d.c. because the field is a permanent magnet. Alternating current measurements are made possible by the incorporation of a rectifier.

Moving-iron type instruments (Fig. 21.2)

These meters are either **attraction-** or **repulsion-**type instruments.

The attraction-type instrument consists of a soft iron plate attached to a spindle which is mounted between jewelled bearings. A coil is wound on a former which has a rectangular core. When current flows through the coil, the plate is attracted towards the core and a pointer attached to the spindle moves to indicate the amount of current flow. The movement is not proportional to the current and the scale graduations are cramped at the lower end, i.e. the scale is **non-linear**. This instrument measures a.c. or d.c. quantities.

The controlling torque is by gravity acting on the control weight. Damping is achieved by means of a light aluminium piston moving in a curved cylinder. The weight B acts as a counterbalance to the combined weight of the plate, pointer and moving part of the damping device.

A repulsion-type instrument employs two iron plates positioned in a solenoid. One of these is fixed and the other is attached to a spindle and pointer. When current flows through the solenoid, the plates are magnetised in the same sense and a force of repulsion between them causes the pointer to move.

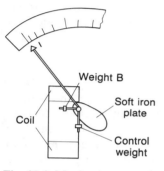

Fig. 21.2 Moving-iron type instrument

21.3 User precautions

It is important that all measuring instruments are treated with respect and handled correctly if damage to the instrument is to be avoided. The following precautions are recommended when using instruments but must not be regarded as a complete list.

Handling

Care is necessary to prevent the instrument from being dropped or bumped. Rough handling causes damage to the meter movement and consequently to the accuracy of the instrument.

Location

The instrument should be placed on a flat, firm surface which is free from vibration and external magnetic fields and once in position should not be moved unnecessarily.

Range selection

The required range must be decided upon before connecting the instrument in circuit, with a high range being selected if any doubt exists.

Zero check

Mechanical and electrical zero should be checked before use and adjusted where necessary.

Connections

It is essential that the instrument is connected in circuit as appropriate to the measurement being made. All electrical connections must be clean and tight.

Batteries

Instruments requiring batteries for their operation should have their batteries inspected periodically for signs of corrosion. When a resistance range is selected on some models of multimeter, the battery supplies current even when an actual test is not being made.

21.4 Insulation testing

Many insulating materials are designed to isolate and protect various conductive elements. An indication of the effectiveness of such insulating materials is achievable by subjecting them to a **voltage stress** and measuring their **insulation resistance**. The instrument used for this purpose is the **megohmeter** or **megger**.

21.5 The cathode-ray oscilloscope

A cathode-ray tube is shown schematically in Fig. 21.3. Electrons emitted from the cathode are focused into a thin stream under the effect of a voltage applied to the first anode and then accelerated to a fluorescent screen. When the electrons strike the screen, they produce a luminous spot visible from outside the tube. The intensity of this spot is controlled by the grid voltage. The electron stream is deflected by a magnetic field produced by coils positioned around the neck of the tube or, more usually, by electrostatic means. The stream passes through two sets of plates set at right-angles to each other. The X plates provide horizontal deflection and the Y plates vertical deflection.

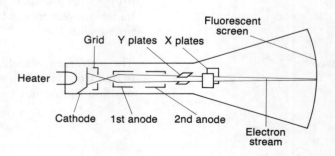

Fig. Fig. 21.3 Cathode-ray tube

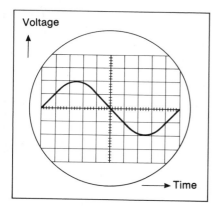

Fig. 21.4

An oscilloscope is effectively a visual voltmeter. A time-base circuit is connected to the X plates causing the spot to move horizontally across the screen. At most of the speeds involved, this spot appears as a continuous line. The input signal to be viewed is usually connected to the Y plates, having first been amplified by an amplifier inside the oscilloscope.

A typical oscilloscope display is illustrated in Fig. 21.4.

The screen is divided into centimetre squares and each square is sub-divided into five parts. The horizontal axis (X axis) represents time and the vertical axis (Y axis) voltage.

Figure 21.4 shows one complete cycle of a sine wave displayed on the screen. The time occupied by one cycle is termed the **periodic time** and by referring to the switch settings on the oscilloscope this time can be calculated.

Assume for Fig. 21.4 that this time is 0.01 seconds. This implies that 100 cycles of this waveform occurs every second, or that the frequency of the signal is 100 hertz.

The relationship between periodic time T and frequency f is

$$f = \frac{1}{T} \text{ Hz.}$$

Some oscilloscopes are known as **dual beam** and produce two electron streams. These are most useful for the simultaneous comparison of two signals.

Exercises

2 Resistors and Ohm's law

2.1 A resistor has four bands coloured yellow, violet, orange and silver. Determine the resistance value of the component.

2.2 A resistor has a value of 100 ohms with a tolerance of 1%. Write down the colours of the four bands painted on the component.

2.3 An e.m.f. of 12 V is applied to a resistive load of 5 kΩ. Determine the current flow in the circuit.

2.4 A circuit has a total resistance of 18 kΩ. An ammeter connected into the circuit indicates a current flow of 2.5 mA. Determine the supply voltage.

2.5 A load draws 80 mA from a 24 V supply. Determine the resistance of the load.

3 Resistors in series

3.1 A resistor coloured orange, white and orange is connected in series with a resistor coloured grey, red and orange. Determine the total resistance of the two components.

3.2 A circuit consists of three resistors connected in series. Their values are 470 Ω, 260 Ω and 70 Ω. If the supply voltage is 24 V, determine the current flow in the circuit.

3.3 Two resistors are connected in series across a 12 V supply. Their values are 56 Ω and 40 Ω. Determine the potential difference across the 40 Ω resistor.

3.4 An electrical supply of 20 V has an internal resistance of 80 Ω. Determine the terminal voltage when the load resistance is 420 Ω.

4 Resistors in parallel

4.1 A circuit consists of three resistors connected in parallel. Their values are 4 Ω, 6 Ω and 12 Ω. If 12 V is applied to the circuit, determine (*a*) the total circuit resistance, (*b*) the current flowing through each resistor and (*c*) the current drawn from the supply.

4.2 Determine the total resistance of three resistors connected in parallel whose values are 18 MΩ, 12 MΩ and 9 MΩ.

4.3 Two resistors of 500 Ω and 125 Ω connected in parallel are connected in series with a resistor of 75 Ω. Calculate the total resistance of the combination.

5 Electrical power

5.1 Calculate the power dissipated by a bulb which uses 1200 joules in 40 seconds.

5.2 A current of 4 A flows in a bulb having an effective resistance of 3 Ω. Calculate the power dissipated by the bulb.

5.3 A battery maintains a constant voltage of 12 V when supplying a headlamp rated at 48 W. The battery has a capacity of 180 A h. Calculate how long it will take to reduce the charge on the battery to three-quarters of full charge.

6 Magnetism and electromagnetism

6.1 20 V a.c. is applied to the primary winding of a 1 to 9 step-up transformer. Determine the output voltage.

6.2 When 15 V a.c. is applied to the primary winding of a transformer, the output voltage is measured as 120 V a.c. If the secondary winding has 4000 turns, determine the number of turns of the primary winding.

6.3 Two 100 μH inductors are connected in parallel. Determine the total inductance of the combination.

7 Capacitance

7.1 A 15 μF capacitor is connected in series with a 30 μF capacitor. If these components are connected in parallel with a 60 μF capacitor, determine the total capacitance of the network.

7.2 The time constant for a capacitor connected in series with a resistor is measured as 8 seconds. Calculate the value of the resistor when the capacitor is 160 μF.

12 d.c. generators

12.1 The armature of a d.c. generator has a resistance of 0.2 Ω. If on full load the terminal voltage falls by 5.5 V, calculate the armature current at full load.

12.2 The generated e.m.f. of a generator is 240 V. The machine produces a terminal voltage of 230 V when delivering 80 A to the load. Calculate the armature resistance.

13 The charging sub-system

13.1 In the charging circuit shown, a six-cell lead-acid battery is charged by a constant voltage generator producing 18 V. If the total resistance of the charging circuit is 0.6 Ω, calculate the charging current when the terminal voltage of each cell is (*a*) 1.8 V and (*b*) 2.5 V.

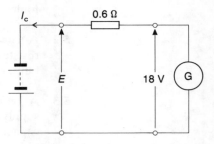

Fig. E13.1

15 d.c. motors

15.1 The armature of a d.c. motor has a resistance of 0.1 Ω. When the applied voltage is 235 V, the generated back e.m.f. is 226 V. Calculate the armature current drawn from the source.

15.2 A change in load for the d.c. motor in Exercise 15.1 causes the armature current to increase by 150.67%. Calculate the percentage fall in motor speed.

16 d.c. machines – further considerations

16.1 A generator provides 15 kW of electrical power. If the generator is 85% efficient, calculate the mechanical input power in kilowatts required to drive the machine.

16.2 Determine the power rating in watts of a motor rated at 14 h.p.

16.3 An engine develops 12 h.p. when driving a generator. Calculate the power losses in horsepower of the generator when the power dissipated in an electrical load is 8.36 kW.

17 Alternating current and a.c. alternators

17.1 Determine the peak value of an alternating voltage of 141.4 V r.m.s.

17.2 The current in an a.c. circuit is measured as 183.82 A r.m.s. Determine the peak value of the current.

18 Reactance

18.1 Calculate the reactance of a coil of inductance 3.95 mH when it is connected to a 10 kHz supply.

18.2 A coil has a reactance of 558 Ω when connected in a circuit of frequency 5 kHz. Determine the inductance of the coil.

18.3 A 15 pF capacitor is connected to a circuit of frequency 20 MHz. Determine the capacitive reactance.

18.4 Calculate the current taken by a 30 μF capacitor when it is connected to a 240 V, 50 Hz supply.

21 Measuring instruments

21.1 An alternating current completes eight cycles in 200 μs. Determine the frequency.

21.2 Determine the periodic times for frequencies of (a) 200 Hz and (b) 5 MHz.

Answers to exercises

2.1 47 kΩ ± 10% **2.2** Brown, black, brown, brown
2.3 2.4 mA **2.4** 45 V **2.5** 300 Ω

3.1 121 kΩ **3.2** 30 mA **3.3** 5 V **3.4** 16.8 V

4.1 (*a*) 2 Ω (*b*) 3 A, 2 A, 1 A (*c*) 6 A **4.2** 4 MΩ **4.3** 175 Ω

5.1 30 W **5.2** 48 W **5.3** 11.25 hours

6.1 180 V **6.2** 500 **6.3** 50 µH

7.1 70 µF **7.2** 50 kΩ

12.1 27.5 A **12.2** 0.125 Ω

13.1 (*a*) 12 A (*b*) 5 A

15.1 90 A **15.2** 6%

16.1 17.65 kW **16.2** 10.44 kW **16.3** 0.79 h.p.

17.1 200 V **17.2** 260 A

18.1 248.2 Ω **18.2** 17.76 mH **18.3** 530.52 Ω **18.4** 2.26 A

21.1 40 kHz **21.2** (*a*) 5 ms (*b*) 0.2 µs

Index